The Electron, Proton, and Neutron

The discoveries of twentieth-century physics have played a decisive influence in the thinking of modern man. With the exploration of the atom has come a staggering vision of the complexity of the physical world. At the same time, the pursuit of such still puzzling factors as the elusive electron, the mysterious muon, and the strange multiplicity of baryons has given us but a hint of how much is yet to be known. In the concluding volume of his celebrated UNDERSTANDING PHYSICS, Isaac Asimov brilliantly conveys this vast new knowledge and awesome sense of challenge. Treating the most advanced ideas with superb clarity, his book is the perfect supplement to the student's formal textbook, as well as offering invaluable illumination to the general reader.

ABOUT THE AUTHOR: ISAAC ASIMOV is generally regarded as one of this country's leading writers of science and science fiction. He obtained his Ph.D. in chemistry from Columbia University; currently he is Associate Professor of Biochemistry at Boston University School of Medicine. He is the author of over sixty books, including *The Chemicals of Life, The Genetic Code, The Human Body, I, Robot,* and *The Wellsprings of Life,* all available in Signet editions.

UNDERSTANDING PHYSICS

Volume III

The Electron, Proton, and Neutron

ISAAC ASIMOV

A SIGNET BOOK
NEW AMERICAN LIBRARY
TIMES MIRROR

MENTOR TRADEMARK REG. U.S. PAT. OFF. AND FOREIGN COUNTRIES
REGISTERED TRADEMARK—MARCA REGISTRADA
HECHO EN CHICAGO, U.S.A.

SIGNET, SIGNET CLASSICS, MENTOR, PLUME AND MERIDIAN BOOKS
are published in the United States by
The New American Library, Inc.,
*1301 Avenue of the Americas, New York, New York 10019,
in Canada by* The New American Library of Canada Ltd.,
81 Mack Avenue, Scarbrough, 704, Ontario

FIRST PRINTING, APRIL, 1969

4 5 6 7 8 9 10 11 12

PRINTED IN THE UNITED STATES OF AMERICA

TABLE OF CONTENTS

UNDERSTANDING PHYSICS

The Electron, Proton, and Neutron

CHAPTER 1

The Atom

In the first two volumes of this book I dealt with those aspects of physics in which the fine structure of matter could be ignored.

I discussed gravitation, for instance, since any sphere possessing the mass of the earth would exhibit the gravitational effect of the earth regardless of the type of matter of which it was composed. Furthermore, the question of the ultimate structure of the finest particles of matter need not be considered in working out the laws governing the gravitational interaction of bodies.

The same is true of the laws of motion. A brick moves as a unit and we need not be concerned with the composition of the brick in studying its motion. We can study many phases of the electric charge of a pith ball, or the magnetic field of a magnet, and derive useful laws governing electromagnetic effects without probing into the submicroscopic structure of the magnet. Even heat can be considered a subtle fluid pouring from one object to another, and the laws of thermodynamics can be deduced from this sort of concept.

Yet in the course of these two volumes, it has become plain, every once in a while, that a deeper understanding of phenomena than that offered by the everyday world of ordinary objects can be achieved if we burrow down into the ultra-small.

For instance, the properties of gases are best understood if we consider them to be composed of tiny particles (see page

I–200).* Temperature can best be understood if it is considered to represent the average kinetic energy of tiny particles of matter in motion (see page I–205). Energy, as well as matter, seems to make more sense if it is considered as consisting of tiny particles (see page II–130).

In this third volume, therefore, I will go into the fine structure of matter and energy in some detail. I will try to show how physical experimentation revealed a world beyond the direct reach of our senses, and how knowledge of that world has, in turn, lent more meaning to the ordinary world we can sense directly.

Origin of Atomism

The notion of *atomism* (a name we can give to the theory that matter is composed of tiny particles) arose first among the Greeks—not as a result of experiment, but as a result of philosophic deduction.

Atomism is by no means self-evident. If we can trust our senses, most types of matter seem "all one piece." A sheet of paper or a drop of water does not seem to be composed of particles.

This, however, is not conclusive. The sand making up a beach, if viewed from a distance, seems to be all one piece. It is only upon a close view that we can make out the small grains of which the sand actually consists. Perhaps, then, paper or water is made up of particles too small to see.

One way of testing the matter is by considering the divisibility of a substance. If you had a handful of sand and ignored the evidence of your eyes, seeking instead some other criterion of atomism, you might begin by dividing the handful into two portions with your finger, dividing each of those into two still smaller portions, and so on. Eventually, you would find yourself in possession of a portion so small as to consist of a single grain, and this final portion could no longer be divided by finger. We might consider atomism, then, as implying that matter cannot be divided and subdivided indefinitely. At some point, a unit no longer divisible by a method that had sufficed earlier must be obtained.

If this is true of paper or water, however, the ultimate pieces are far too small to see. No limiting indivisible unit in matter generally can be directly sensed. Can the existence of such ultra-small units be deduced by reason alone?

---

* References to material in the first two volumes will be indicated by phrases such as "(see page I–123)" or "(see page II–123)." References to material in this third volume will be indicated by a simple "(see page 123)."

The opportunity arose in the fifth century B.C. with the paradoxes raised by Zeno of Elea. Zeno pointed out that one could reach conclusions by reason that seemed to contradict the evidence of the senses and that it was necessary, therefore, to search for flaws either in the process of reasoning or in sense-perception. His most famous paradox is called "Achilles and the Tortoise."

Suppose that the Greek hero Achilles, renowned for his fleetness of foot, can run ten times as fast as a tortoise. The tortoise is now given a hundred-yard headstart, and the two race. By the time Achilles covers the hundred yards that separated him from the tortoise, the tortoise has moved forward ten yards. When Achilles makes up that ten yards, the tortoise has moved forward one yard; when Achilles bounds across that one yard, the tortoise has moved forward a tenth of a yard, and so on. By this line of reasoning, it seems clear that Achilles can never catch up to the tortoise, who always remains ahead (though by a smaller and smaller margin). And yet, in any such race we certainly know that Achilles will, in actual fact, overtake and pass the tortoise.

Nowadays, mathematicians understand that the successive margins of the tortoise—100 yards, 10, 1, 0.1, and so on—make up a "converging series." A converging series may have an infinite number of terms, but these will nevertheless come to a definite and finite sum. Thus, the converging series consisting of $100 + 10 + 1 + 0.1 + 0.01$, etc., has the finite sum of $111\frac{1}{9}$. This means that after Achilles has run $111\frac{1}{9}$ yards he will be exactly even with the tortoise and that thereafter he will forge ahead.

The Greeks, however, knew nothing of converging series and had to find other reasons for reconciling Zeno's argument with the facts of life. One way out was to consider that Zeno divided the distance between Achilles and the tortoise into smaller and smaller portions with no indication that any portion was so small that it could no longer be divided into anything smaller.

Perhaps that was not the way the universe worked. Perhaps there were units so small that they could be divided no further. If this notion of limited divisibility was adopted, perhaps Zeno's paradoxes, based on unlimited divisibility, might disappear.

It may have been reasoning like this that led some Greek philosophers to suggest that the universe was made up of tiny particles that were themselves indivisible. The most prominent of these philosophers was Democritus of Abdera, who advanced his theories about 430 B.C. He called these ultimate particles "atomos," from a Greek word meaning "indivisible," and it is from this that our word, *atom*, is derived.

Democritus went on to interpret the universe in atomic terms and came up with a number of suggestions that sound quite modern. However, it all rested on pure reasoning. He could suggest no evidence for the existence of atoms other than "this is the way it must be."

Other Greek philosophers of the time could offer arguments for the nonexistence of atoms on a "this is the way it must be" basis. On the whole, most ancient philosophers went along with non-atomism, and Democritus' views were buried under the weight of adverse opinions. In fact, so little worthy of attention was Democritus considered that his works were infrequently copied, and none of his voluminous writings have survived into modern times. All we know of him are offhand remarks in the works of such philosophers as have survived, just about all of whom are non-atomists and therefore mention Democritus' views only disparagingly.

Nevertheless, his views, however crushed, did not altogether die. Epicurus of Samos (341–270 B.C.), who began teaching in Athens in 306 B.C., incorporated the atomism of Democritus into his philosophic system. Although Epicureanism proved quite influential in the next few centuries, none of the works of Epicurus have survived, either.

Fortunately, however, the works of one Epicurean philosopher have survived. The Roman poet Lucretius (96?–55 B.C.) wrote a long poem *De rerum natura* ("On the Nature of Things") in which he interpreted the universe in Epicurean fashion, making use of an atomistic viewpoint. *One* copy of this poem survived, and when printing was invented in the fifteenth century, it was one of the first of the ancient classics to be printed.

As modern science came to birth, then, atomistic views were present for the plucking. A French philosopher, Pierre Gassendi (1592–1655), adopted the Epicurean views of Lucretius and was influential in spreading the doctrine of atomism.

One of those who came under the influence of Gassendi was the English scientist Robert Boyle (1627–1691), and with him atomism enters a new phase; it is no longer a matter of philosophy and deduction, but rather one of experiment and observation.

The Chemical Elements

Boyle studied air and found that it could be compressed or expanded (see page I–145). In other words, the volume of a gas could be changed without changing its mass. It is difficult to imagine how this could happen if matter were really continuous. Rubber can be stretched, to be sure, but as a rubber band grows longer, it also grows thinner; the volume is not perceptibly altered.

The behavior of air is much more like that of a sponge, which can be compressed in all directions or pulled apart in all directions—its volume considerably changed without a change in mass. In the case of a sponge, the explanation involves the numerous air-filled cavities. The sponge can be compressed because the cavities can be forced closed by squeezing out the air. It can expand once more if air is allowed to re-enter the cavities.

Is it possible, then, that invisible cavities exist in air itself, cavities which can be squeezed closed when air is compressed and made to open wide when it is expanded? This can, in fact, be visualized if it is supposed that air is made up of a myriad of ultra-tiny particles separated by utter emptiness. Compression would involve the movement closer together of these particles; expansion, the movement farther apart. Volume would change while mass (which would depend merely on the number of particles and not on their distance apart) would not. Other properties of gases could likewise be handily explained by atomistic reasoning.

Atomism could surely be transferred from gases to solids and liquids, since the latter can easily be converted, through heat, to gases or vapors. Thus boiling water (or even water standing at ordinary temperatures) is converted into water vapor, a gas that is much less dense than liquid water. The vapor can be condensed into liquid water once more. To explain this, we might suppose that water consists of atoms packed very closely. (From the fact that water, in common with other liquids and solids, cannot be compressed appreciably by the forces that compress gases easily, we might even suppose that the atoms in liquids and solids are in contact.) When a liquid evaporates, the molecules are pulled apart; when a gas is condensed, its molecules are forced together.

Despite reasoning of this sort, the world of science found it difficult to accept atomism. The philosophical difficulty of dealing with objects infinitesimal in size and undetectable by any device then known was too great.

What finally established atomism firmly was the slow gathering of a quantity of chemical evidence in its favor. To describe this, I will begin with the concept of an *element*.

The Greeks were the first to speculate on the nature of the fundamental substance or substances (elements) of which the universe was composed. Their speculations, in the absence of actual chemical experimentation, were really only guesses, but since the Greek thinkers were highly intelligent men, they produced extraordinarily sensible guesses.

Aristotle (384–322 B.C.) summarized Greek labors in this direction by listing four elements in the world itself: earth, water, air, and fire; and a fifth element, making up the incorruptible heavens above: ether (see page I–6). If, for the four earthly elements, we used the closely allied words: solid, liquid, gas, and energy—we would see that the guesses were indeed sensible.

For two thousand years, the Greek notion of the four earthly elements survived. By 1600, however, the notion of experimentation was beginning to preoccupy scientists, thanks very largely to the work of Galileo Galilei (1564–1642). An element, it seemed, should be defined in experimental terms; it should be defined as something that was capable of doing something (or incapable of doing something) rather than of merely being something. It needs what is now called an *operational definition*.

In 1661, Robert Boyle wrote a book called *The Sceptical Chymist* in which he explained his notion of an element. If an element was indeed one of the simple substances out of which the universe was composed, then it should certainly not be capable of being broken down to still simpler substances or of being produced through the union of still simpler substances. As soon as a substance was broken into simpler substances, it was, at once and forever, not an element.

Since earth is easily separated into different substances, earth is not an element. A century after Boyle's time, air and water were separated into simpler substances and were thus shown not to be elements. As for fire, chemists came to realize that it was not matter at all but a form of energy, and therefore it fell outside the world of elements altogether.

For a considerable time after Boyle, chemists could never know for sure whether a given substance was an element, since one could never tell when new experimental techniques might be developed which would make it possible to break down a previously untouchable substance.

As an example, the substance known as "lime" (or, in Latin, *calx*) had to be considered an element throughout the eighteenth century, since nothing chemists could do would break it down to simpler substances. There were reasons, though, for suspecting that it consisted of an unknown metal combined with a gas, oxygen. However, this was not shown to be fact until 1808, when the English chemist Humphry Davy (1778–1829), succeeded in decomposing lime and isolating a new metallic element, *calcium* (from lime's Latin name). For this, however, he had had to make use of a current of electricity, a new discovery then.

For ease in referring to the elements, a *chemical symbol* for each was introduced in 1814 by the Swedish chemist Jöns Jakob Berzelius (1779–1848). Essentially, these symbols consist of the initial letter of the Latin name (usually, but not always, very similar to the English name) plus (again usually, but not always) a second letter from the body of the name. The symbols used are in almost every case so logical that after very little practice their meanings come to offer no difficulty whatever.

During the course of the nineteenth century, chemists grew to understand the nature of elements, and by the early decades of the twentieth century, elements could be defined with remarkable precision. The manner in which this came about will be described later in this book, but meanwhile, I will list (alphabetically), together with the chemical symbols for each, the substances now recognized as elements (see Table I).

The Modern Atomic Theory

Of course, not all substances found in nature are elements. Most substances are composed of two or more elements, not merely mixed but intimately joined in such a fashion that the final substance has properties of its own that are not necessarily similar to those of any of the elements making it up. Such a substance, formed of an intimate union of elements, is called a *compound*

In the latter part of the eighteenth century, chemists forming their compounds began to study more than the merely qualitative nature of the products formed in their reactions. It was no longer enough merely to note that a gas had bubbled off or that a floc-culent material of a certain color had settled to the bottom of a container. Chemists took to measurement—to determining the actual quantity of substances consumed and produced in their reactions.

The most prominent in establishing this new trend was the French chemist Antoine Laurent Lavoisier (1743–1794), who for this and other services is commonly called "the father of modern chemistry." Lavoisier gathered enough data by 1789 to be able to maintain that in any chemical reaction in a closed system (that is, one from which no material substance may escape and into which no material substance may enter) the total mass is left unchanged. This is the *law of conservation of matter*, or the *law of conservation of mass*.

It was an easy step from this to the separate measurement of the mass of each component of a compound. Important work in this respect was done by the French chemist Joseph Louis Proust (1754–1826). He worked, for example, with a certain compound, now called copper carbonate, which is made up of three elements: copper, carbon and oxygen. Proust began with a pure sample of copper carbonate, broke it down into these three elements, and determined the mass of each element separately. He found the elements always present in certain fixed proportions: for every five parts of copper (by weight) there were four parts of oxygen and one part of carbon. This was true for all the samples

TABLE I—*Elements and Their Symbols*

Actinium	Ac	Erbium	Er
Aluminum	Al	Europium	Eu
Americium	Am	Fermium	Fm
Antimony	Sb	Fluorine	F
Argon	Ar	Francium	Fr
Arsenic	As	Gadolinium	Gd
Astatine	At	Gallium	Ga
Barium	Ba	Germanium	Ge
Berkelium	Bk	Gold	Au
Beryllium	Be	Hafnium	Hf
Bismuth	Bi	Helium	He
Boron	B	Holmium	Ho
Bromine	Br	Hydrogen	H
Cadmium	Cd	Indium	In
Calcium	Ca	Iodine	I
Californium	Cf	Iridium	Ir
Carbon	C	Iron	Fe
Cerium	Ce	Krypton	Kr
Cesium	Cs	Lanthanum	La
Chlorine	Cl	Lawrencium	Lw
Chromium	Cr	Lead	Pb
Cobalt	Co	Lithium	Li
Copper	Cu	Lutetium	Lu
Curium	Cm	Magnesium	Mg
Dysprosium	Dy	Manganese	Mn
Einsteinium	Es	Mendelevium	Md

of copper carbonate he tested, no matter how they were prepared. It was as though elements would fit together in certain definite proportions and no other.

Proust found this was true for other compounds that he tested, and he announced his finding in 1797. It is sometimes called *Proust's law*, sometimes *the law of fixed proportions*, and sometimes *the law of definite proportions*.

It is the law of fixed proportions that forced the concept of atomism to arise out of purely chemical considerations. Suppose that copper consists of tiny copper atoms; oxygen, of oxygen atoms; and carbon, of carbon atoms. Suppose further that copper carbonate is formed when a copper atom, an oxygen atom and a carbon atom all join in a tight union. (The truth of the matter is

Mercury	Hg	Samarium	Sm
Molybdenum	Mo	Scandium	Sc
Neodymium	Nd	Selenium	Se
Neon	Ne	Silicon	Si
Neptunium	Np	Silver	Ag
Nickel	Ni	Sodium	Na
Niobium	Nb	Strontium	Sr
Nitrogen	N	Sulfur	S
Nobelium*	No	Tantalum	Ta
Osmium	Os	Technetium	Tc
Oxygen	O	Tellurium	Te
Palladium	Pd	Terbium	Tb
Phosphorus	P	Thallium	Tl
Platinum	Pt	Thorium	Th
Plutonium	Pu	Thulium	Tm
Polonium	Po	Tin	Sn
Potassium	K	Titanium	Ti
Praseodymium	Pr	Tungsten	W
Promethium	Pm	Uranium	U
Protactinium	Pa	Vanadium	V
Radium	Ra	Xenon	Xe
Radon	Rn	Ytterbium	Yb
Rhenium	Re	Yttrium	Y
Rhodium	Rh	Zinc	Zn
Rubidium	Rb	Zirconium	Zr
Ruthenium	Ru		

* Name not yet official

more complicated than this, but right now we are only trying to observe the consequences of an atomistic supposition.) A tight union of atoms, such as that which I am suggesting, is called a *molecule* (from a Latin word meaning "a small mass"). What I am saying, then, is suppose that copper carbonate is made up of molecules, each containing a copper atom, a carbon atom and an oxygen atom.

What, now, if it happened that a copper atom was five times as massive as a carbon atom, and an oxygen atom was four times as massive as a carbon atom? It would then be expected that copper carbonate would have to contain five parts of copper (by weight) to four parts of oxygen to one part of carbon. In order to have 5.1 parts of copper to one part of carbon, or 3.9 parts of oxygen to one part of carbon, we would need to work with fractions of atoms.

But this never happens. Only certain proportions exist within a compound and these cannot be varied through slight amounts in this direction and that. This shows that from Proust's law of fixed proportions we can not only reasonably speak of atoms, but that we must come to the decision that the atoms were indivisible, as Democritus had imagined so many centuries before.

These thoughts occurred, in particular, to an English chemist, John Dalton (1766–1844). Based on the law of fixed proportions and on other generalizations of a similar nature, he advanced the *modern atomic theory* (so called to distinguish it from the ancient one advanced by Democritus) in 1803. Dalton recognized the honor due Democritus, for he carefully kept the ancient philosopher's term "atom."

Dalton could go much further than Democritus, of course. He did not need to confine himself to the statement that atoms existed. From the law of fixed proportions it was quite plain that:

(1) Each element is made up of a number of atoms all with the same fixed mass.

(2) Different elements are distinguished by being made up of atoms of different mass.

(3) Compounds are formed by the union of small numbers of atoms into molecules.*

* As a matter of fact, each of these three statements proved to be wrong, as chemists found when they probed more deeply into the fundamental structure of matter. Nevertheless, for those substances most easily dealt with by early nineteenth-century techniques, they were reasonably correct. Dalton's propositions represent a "first approximation" that served to start investigations in the right direction and made it possible to improve those starting approximations as further data were gathered. In science, it is not all-important to be Right (it

From the law of fixed proportions it is even possible to come to conclusions about the relative mass of the different kinds of atoms. This relative mass is commonly referred to as *atomic weight*.†

For instance, water is made up of hydrogen and oxygen, and in forming water it is found that one part of hydrogen (by weight) combines with eight parts of oxygen. Dalton was convinced that compounds were formed by the union of as few atoms as possible, so he considered a molecule of water to be made up of one atom of hydrogen combined with one atom of oxygen. In that case, it was easy for him to decide that an oxygen atom must be eight times as massive as a hydrogen atom.

This does not tell us what the actual mass of either the oxygen atom or the hydrogen atom is, but it does not represent checkmate by any means. Dalton decided to use the hydrogen atom as a reference because he suspected it to be the lightest atom (and here, as it happens, he proved to be right), and he set its mass arbitrarily equal to 1. On that hydrogen = 1 basis, he could set the mass of the oxygen atom at 8.

But a refinement became necessary at this point. It turned out that at just about the time that Dalton was working out his atomic theory, water was being split up into hydrogen and oxygen by the action of an electric current. When this was done, it was found that for every liter of oxygen evolved, two liters of hydrogen were produced. The ratio (by volume) was two parts of hydrogen to one part of oxygen. It was not long before this was shown to mean that the water molecule was composed of two hydrogen atoms and one oxygen atom (though Dalton himself never accepted this).

The molecule can be represented by a *chemical formula* in which an atom of each element contained is represented by its chemical symbol. Thus, Dalton's conception of the water molecule would be HO. Where more than one atom of a particular element is present in the molecule, the number is indicated by a numerical subscript. Therefore, the molecule of water, as accepted now, would have a formula of H_2O.

---

may even be that there is no way of ever determining what is Right); it is merely necessary to be right enough for the times, and Dalton was every bit of that.

† Weight is not the same as mass (see page 1-53), and it would be more scientifically appropriate to speak of "atomic mass" rather than atomic weight. However, in this case, as in many others, an unfortunate word or phrase has entered the scientific literature and has become so popular and well-known as to be impossible to change. Such things must, with a sigh, be lived with.

Changing one's deductions does not change the nature of the experimental observations. Water remains made up of one part of hydrogen (by weight) to eight parts of oxygen. Under the new interpretation of the molecular structure of water, however, the one oxygen atom in the molecule must now be eight times as massive as both hydrogen atoms taken together and sixteen times as massive, therefore, as a single hydrogen atom. Therefore, if we set the atomic weight of hydrogen arbitrarily equal to 1, the atomic weight of oxygen must be equal to 16.

This system can then be used to leapfrog from element to element. For instance, carbon dioxide is produced when three parts of carbon are combined with eight parts of oxygen (by weight). The molecule of carbon dioxide contains one atom of carbon and two atoms of oxygen (CO_2). This means that one atom of carbon is 3/8 as massive as two atoms of oxygen. Since the atomic weight of oxygen is 16, two atoms of oxygen must have a mass of 32. If the carbon atom has a mass 3/8 times 32, its atomic weight is 12.

The molecule of cyanogen (C_2N_2) contains six parts of carbon (by weight) to seven of nitrogen. The two atoms of carbon have a mass of 24; therefore the two atoms of nitrogen have a mass of 24 times 7/6, or 28, and a single atom of nitrogen has an atomic number of 14.

It would seem from this that atomic numbers can be expressed as integers on a hydrogen = 1 basis, and Dalton was indeed convinced that this was true. However, over the next decades, other chemists, notably Berzelius, made more accurate determinations and found that some atomic weights were not integers at all. The atomic weight of chlorine is approximately 35.5, for instance, and the atomic weight of magnesium is 24.3.

Indeed, even some of the atomic weights that seem integers turn out to be not quite integers if very accurate measurements are made. For instance, the proportions of oxygen and hydrogen in water are not exactly 8 to 1 by weight, but rather 7.94 to 1. This means that if we set the atomic weight of hydrogen arbitrarily equal to 1, then the atomic weight of oxygen is 15.88.

But oxygen combines easily with many elements. Of all the elements readily available to the chemists of the early nineteenth century, oxygen combined most readily with other elements. (It was *chemically active*.) Its readiness to combine made oxygen particularly useful in calculating atomic weights, and to have its atomic weight set at some fractional value meant needless complexity of arithmetical computations. Chemists eventually decided, therefore, to set the atomic weight of oxygen exactly equal to 16.0000 and let that serve as standard. The atomic weight of hydrogen would then be 1.008.

This served satisfactorily for nearly a century. By 1920, however, new facts concerning atoms were learned (see Chapter 8) which made the standard of oxygen = 16.0000 inadequate. However, the standard had become so fixed in the literature and in chemical consciousness that it was difficult to change. In 1961, however, a new and better system was adopted (see page 148) which involved a change so slight that it could be tolerated. By the 1961 system, the atomic weight of oxygen, for instance, is 15.9994.

Of the 103 known elements, 83 occur in the earth's crust to an appreciable extent. In Table II these 83 elements are listed in order of increasing atomic weight, and the atomic weight of each, by the 1961 system, is given. The question of the masses of the 20 remaining elements will be considered in another chapter.

The Periodic Table

By the mid-nineteenth century, two definitions of an element were available. One was Boyle's definition (that of a substance that could not be broken down to two or more still simpler substances), and one was Dalton's definition (that of a substance made up entirely of atoms of a given atomic weight). There was no conflict between the two, for the same list of substances qualified as elements by either definition. However, there was an embarrassment of riches—too many elements for comfort. By the 1860's, more than sixty elements were known.

These came in a wide variety of properties: some were gases at ordinary temperatures, a few were liquids, and most were solids; some were nonmetals, some light metals, some heavy metals, and some semimetals; some were very active, some moderately active, and some quite inactive; some were colored and some were not.

All this was rather upsetting. Scientists must take the universe as they find it, of course, but there is a deep-seated faith (no other word will suffice) dating back to Greek times that the universe exhibits order and is basically simple. Whenever any facet seems to grow tangled and complex, scientists can't help searching for some underlying order that may be eluding them.

Attempts were made in the mid-nineteenth century to find such an order among the elements. As the tables of atomic weights

TABLE II—The Atomic Weights of Elements

Element	Atomic Weight	Element	Atomic Weight
Hydrogen	1.00797	Ruthenium	101.07
Helium	4.0026	Rhodium	102.905
Lithium	6.939	Palladium	105.4
Beryllium	9.0122	Silver	107.870
Boron	10.811	Cadmium	112.40
Carbon	12.01115	Indium	114.82
Nitrogen	14.0067	Tin	118.69
Oxygen	15.9994	Antimony	121.75
Fluorine	18.9984	Iodine	126.9044
Neon	20.183	Tellurium	127.60
Sodium	22.9898	Xenon	131.30
Magnesium	24.312	Cesium	132.905
Aluminum	26.9815	Barium	137.34
Silicon	28.086	Lanthanum	138.91
Phosphorus	30.9738	Cerium	140.12
Sulfur	32.064	Praseodymium	140.907
Chlorine	35.453	Neodymium	144.24
Potassium	39.102	Samarium	150.35
Argon	39.948	Europium	151.96
Calcium	40.08	Gadolinium	157.25
Scandium	44.956	Terbium	158.924
Titanium	47.90	Dysprosium	162.50
Vanadium	50.942	Holmium	164.930
Chromium	51.996	Erbium	167.26
Manganese	54.9380	Thulium	168.934
Iron	55.847	Ytterbium	173.04
Nickel	58.71	Lutetium	174.97
Cobalt	58.9332	Hafnium	178.49
Copper	63.54	Tantalum	180.948
Zinc	65.37	Tungsten	183.85
Gallium	69.72	Rhenium	186.2
Germanium	72.59	Osmium	190.2
Arsenic	74.9216	Iridium	192.2
Selenium	78.96	Platinum	195.09
Bromine	79.909	Gold	196.967
Krypton	83.80	Mercury	200.59
Rubidium	85.47	Thallium	204.37
Strontium	87.62	Lead	207.19
Yttrium	88.905	Bismuth	208.980
Zirconium	91.22	Thorium	232.038
Niobium	92.906	Uranium	238.03
Molybdenum	95.94		

grew more and more accurate and as the concept of atomic weight became clearer to chemists generally, it began to seem logical to arrange the elements in order of increasing atomic weight (as in Table II) and see what could be done with that.

Several efforts of this sort failed, but one succeeded. The success was scored in 1869 by a Russian chemist, Dmitri Ivanovich Mendeleev (1834–1907). Having listed the elements in order of atomic weight, he then arranged them in a table of rows and columns, in such a fashion that elements of similar properties fell into the same column (or row, depending on how the table was oriented). As one went along the table of elements, properties of a certain kind would turn up after fixed periods. For this reason, Mendeleev's product was a *periodic table.*

Difficulties arose out of the fact that the list of elements, extensive as it was, was still incomplete. In order to arrange the known elements in such a way that those of similar properties fell into the same column, Mendeleev found it necessary to leave gaps. These gaps, he announced in 1871, must contain elements not as yet discovered. He announced the properties of the missing elements in some detail, judging these by comparing them with the elements in the same column, above and below the gap, and taking intermediate values.

Within fifteen years, all three elements predicted by Mendeleev were discovered, and their properties were found to be precisely those he had predicted. As a result, the periodic table was, by the 1880's, accepted as a valid guide to order within the jungle of elements, and it has never been abandoned since. Indeed, later discoveries (see page 64) have served merely to strengthen it and increase its value. Mendeleev's discovery had been merely empirical—that is, the periodic table had been found to work, but no reason for its working was known. The twentieth century was to supply the reason.

Table III is a version of the periodic table, as presently accepted. The elements are arranged in order of atomic weight (with three minor exceptions shortly to be mentioned) and are numbered in order from 1 to 103. The significance of this "atomic number" will be discussed on page 64.

If you compare Table III with Table II, you will find that in order to put the elements into the proper rows, three pairs of elements must be placed out of order. Element 18 (argon) though lower in number than element 19 (potassium) has a higher atomic weight. Again, element 27 (cobalt) has a higher atomic weight than element 28 (nickel), while element 52 (tellurium) has a higher atomic weight than element 53 (iodine). In each case the difference in atomic weight is quite small and nineteenth century chemists tended to ignore these few and minor exceptions to the general rule. The twentieth century, however, was to find these exceptions particularly significant (see page 64).

The periodic table contains a number of closely-knit families of elements, with many similarities among their properties. For instance, elements 2, 10, 18, 36, 54 and 86 (helium, neon, argon, krypton, xenon and radon) are the *inert gases*, so called because of their small tendency to react with other substances. Until 1962, in fact, it was thought that none of them underwent any chemical

TABLE III—*The Periodic Table*

1 Hydrogen (H) 1.008								
3 Lithium (Li) 6.939	4 Beryllium (Be) 9.012							
11 Sodium (Na) 22.990	12 Magnesium (Mg) 24.312							
19 Potassium (K) 39.102	20 Calcium (Ca) 40.08	21 Scandium (Sc) 44.956	22 Titanium (Ti) 47.90	23 Vanadium (V) 50.942	24 Chromium (Cr) 51.996	25 Manganese (Mn) 54.938	26 Iron (Fe) 55.847	27 Cobalt (Co) 58.933
37 Rubidium (Rb) 85.47	38 Strontium (Sr) 87.62	39 Yttrium (Y) 88.905	40 Zirconium (Zr) 91.22	41 Niobium (Nb) 92.906	42 Molybdenum (Mo) 95.94	43* Technetium (Tc) 98.91	44 Ruthenium (Ru) 101.07	45 Rhodium (Rh) 102.905
55 Cesium (Cs) 132.905	56 Barium (Ba) 137.34	57 Lanthanum (La) 138.91	58 Cerium (Ce) 140.12	59 Praseodymium (Pr) 140.907	60 Neodymium (Nd) 144.24	61* Promethium (Pm) 145	62 Samarium (Sm) 150.35	63 Europium (Eu) 151.96
			72 Hafnium (Hf) 178.49	73 Tantalum (Ta) 180.948	74 Tungsten (W) 183.85	75 Rhenium (Re) 186.2	76 Osmium (Os) 190.2	77 Iridium (Ir) 192.2
87* Francium (Fr) 223	89* Radium (Ra) 226.05	89* Actinium (Ac) 227	90* Thorium (Th) 232.038	91* Protactinium (Pa) 231	92* Uranium (U) 238.03	93* Neptunium (Np) 237	94* Plutonium (Pu) 242	95* Americium (Am) 243

reactions at all. Since 1962, it has come to be realized that at least three of them, krypton, xenon and radon, will take part in chemical reactions with fluorine.

Again, elements 9, 17, 35, 53, and 85 (fluorine, chlorine, bromine, iodine and astatine) are the *halogens* (from Greek words meaning "salt-formers"). These are active nonmetals that get their family name from the fact that one of them, chlorine, combines with sodium to form ordinary table salt, while the others combine with sodium to form compounds quite similar to salt.

Elements 3, 11, 19, 37, 55, and 87 (lithium, sodium, potassium, rubidium, cesium and francium) are soft, easily melted, very

								2 Helium (He) 4.003
			5 Boron (B) 10.811	6 Carbon (C) 12.011	7 Nitrogen (N) 14.007	8 Oxygen (O) 15.999	9 Fluorine (F) 18.998	10 Neon (Ne) 20.183
			13 Aluminum (Al) 26.982	14 Silicon (Si) 28.086	15 Phosphorus (P) 30.974	16 Sulfur (S) 32.064	17 Chlorine (Cl) 35.453	18 Argon (A) 39.948
28 Nickel (Ni) 58.71	29 Copper (Cu) 63.54	30 Zinc (Zn) 65.37	31 Gallium (Ga) 69.72	32 Germanium (Ge) 72.59	33 Arsenic (As) 74.922	34 Selenium (Se) 78.96	35 Bromine (Br) 79.909	36 Krypton (Kr) 83.80
46 Palladium (Pd) 106.4	47 Silver (Ag) 107.870	48 Cadmium (Cd) 112.40	49 Indium (In) 114.82	50 Tin (Sn) 118.69	51 Antimony (Sb) 121.75	52 Tellurium (Te) 127.60	53 Iodine (I) 126.904	54 Xenon (Xe) 131.30
64 Gadolinium (Gd) 157.25	65 Terbium (Tb) 158.924	66 Dysprosium (Dy) 162.50	67 Holmium (Ho) 164.930	68 Erbium (Er) 167.26	69 Thulium (Tm) 168.934	70 Ytterbium (Yb) 173.04	71 Lutetium (Lu) 174.97	
78 Platinum (Pt) 195.09	79 Gold (Au) 196.967	80 Mercury (Hg) 200.59	81 Thallium (Tl) 204.37	82 Lead (Pb) 207.19	83 Bismuth (Bi) 208.98	84° Polonium (Po) 210	85° Astatine (At) 210	86° Radon (Rn) 222
96° Curium (Cm) 244	97° Berkelium (Bk) 245	98° Californium (Cf) 246	99° Einsteinium (Es) 253	100° Fermium (Fm) 255	101° Mendelevium (Md) 256	102° Nobelium (No) 255	103° Lawrencium (Lw) 257	

active *alkali metals.* The word "alkali" is from an Arabic phrase meaning "ash." It was from the ashes of certain plants that the original "alkalis," soda and potash ("pot-ash") were derived. From these, sodium and potassium, the first alkali metals to be discovered, were obtained by Davy.

Elements 4, 12, 20, 38, 56, and 88 (beryllium, magnesium, calcium, strontium, barium, and radium) are harder, less easily melted, and less active than the alkali metals. They are the *alkaline earth metals.* (An "earth" is an old-fashioned name given to oxides that are insoluble in water and resistant to change under the influence of heat. Two such earths, lime and magnesia, had certain properties resembling those of soda and potash and were therefore called the "alkaline earths." It was from lime and magnesia that Davy obtained calcium and magnesium, the first two alkaline earth metals to be discovered.)

Elements 57 to 71 inclusive form a closely related family of metals that were originally called the *rare earth elements* but have now come to be called the *lanthanides,* from the first element of the group, lanthanum. Elements 89 to 103 inclusive are the *actinides* from actinium, first element of that group.

Other families also exist within the periodic table, but those I have listed are the best known and the most frequently referred to by the family name.

The Reality of Atoms

Once we have the atomic weight, it is easy to see what one means by *molecular weight:* It is the sum of the atomic weights of the atoms making up a molecule. Let us start, for instance, with oxygen, atomic weight 16, and hydrogen, atomic weight 1.*

There is strong chemical evidence to the effect that under ordinary conditions elementary oxygen and hydrogen do not occur as single, separate atoms. Rather, two atoms combine to form a stable molecule, and the gas consists of these two-atom molecules. For this reason, the chemical formulas for gaseous oxygen and gaseous hydrogen are, respectively, O_2 and H_2. If O and H are written, they refer to individual oxygen and hydrogen atoms. You can see, then, that the molecular weight of oxygen is 32 and that of hydrogen is 2.

Again, consider ozone, a form of oxygen in which the mole-

---

* It is often convenient to make use of approximate atomic weights, rounding off the actual value to the nearest integer, or one decimal place at most. When more than that is needed, more than that will be used.

cules are made up of three atoms apiece (O_3). Its molecular weight is 48. That of water (H_2O) is 18. Then, since the atomic weight of carbon is 12, the molecular weight of carbon dioxide (CO_2) is 44.

It is useful for a chemist to consider a quantity of substance with a mass equal to its molecular weight in grams. In other words, he may deal with 2 grams of hydrogen, 32 grams of oxygen, 18 grams of water, or 44 grams of carbon dioxide. Such a mass is the *gram-molecular weight*, which is often spoken of, in abbreviated form, as a *mole*. We can say that a mole of carbon dioxide has a mass of 44 grams, while a mole of ozone has a mass of 48 grams.

Sometimes elements do exist in the form of single, separate atoms. This is true of the inert gases such as helium and argon, for instance. Solid elements, such as carbon and sodium, are for convenience sake often considered to be made up of single-atom units. There we can speak of a *gram-atomic weight*. Since the atomic weight of helium is 4 and that of sodium is 23, the gram-atomic weight of helium is 4 grams and that of sodium is 23 grams. Often, the abbreviated form "mole" is used to cover both gram-molecular weights and gram-atomic weights.

The convenience of the mole in chemical calculations stems from a point first grasped in 1811 by the Italian chemist Amedeo Avogadro (1776–1856) and is therefore called *Avogadro's hypothesis*. Expressed in modern terms, this states: Equal volumes of all gases contain equal numbers of molecules under conditions of fixed temperature and pressure.

In later years, this was found to be correct, at least as a first approximation.

A mole of hydrogen (2 grams) at ordinary air pressure and at a temperature of 0°C. takes up a volume of approximately 22.4 liters. A mole of oxygen (32 grams) is sixteen times as massive as a mole of hydrogen but is made up of molecules that are individually sixteen times as massive as those of hydrogen. Therefore, a mole of oxygen contains the same number of molecules as does a mole of hydrogen. By Avogadro's hypothesis (taken in reverse), this means that 32 grams of oxygen should take up just as much room (22.4 liters) as 2 grams of hydrogen—and they do. The same line of reasoning also applies to other gases.

In short, if we deal with different gases by the mole, we end up with quantities that differ in mass but are equal in volume! The number of molecules present in a mole of gas (any gas) is called *Avogadro's number*.

The equal volume rule holds only for gases, but Avogadro's number is of more widespread use. A mole of any substance— solid or liquid as well as gaseous—contains Avogadro's number of molecules. (Where a substance is made up of individual atoms, as in the case of helium, Avogadro's number of atoms is contained in a gram-atomic weight rather than in a mole, properly speaking, but that is merely a detail.)

If only chemists had known the exact value of Avogadro's number, they could have at once determined the mass of an individual molecule. This would have lent atoms and molecules an air of actuality. As long as they were merely objects "too small to see" and nothing more, they were bound to be considered as merely convenient (and possibly fictitious) ways of explaining chemical reactions. Give an individual atom or molecule a fixed mass, however, find a fixed number in a glass of water or in an ounce of iron, and the small objects begin to seem real.

Unfortunately, it was not for over a half-century after the introduction of the modern atomic theory that the value of Avogadro's number could be determined even approximately. Till then, all that chemists could say was that Avogadro's number was very large.

The break came in 1865. The Scottish physicist James Clerk Maxwell (1831–1879) and the Austrian physicist Ludwig Boltzmann (1844–1906) had worked out the properties of gases by mathematically analyzing the random movements of the tiny atoms or molecules making up that gas (see page I–200). From the equations derived by Maxwell and Boltzmann, it was possible, by making some reasonable suppositions, to calculate what Avogadro's number might be. This was done by a German chemist, J. Loschmidt, and it turned out to be approximately six hundred billion trillion—a large number, indeed.

A number of more accurate methods have been used in the twentieth century for determining the value of Avogadro's number. These have yielded virtual agreement among themselves and have shown Loschmidt's first attempt to be remarkably good. The value of Avogadro's number currently accepted as most nearly accurate is 602,300,000,000,000,000,000,000 or, in exponential notation, 6.023×10^{23}.

If a mole of oxygen gas weighs 32 grams and contains 6.023×10^{23} oxygen molecules, then the individual oxygen molecule must have a mass of 32 divided by 6.023×10^{23}, or about 5.3×10^{-23} grams. Since an oxygen molecule is made up of two oxygen atoms, each of those has a mass of about 2.65×10^{-23} grams. If the mass of the oxygen atom is known, that of all the other atoms can be calculated from the table of atomic weights.

For instance, since the atomic weight of hydrogen is about 1/16 that of oxygen, the mass of the hydrogen atom must be about 1/16 that of the oxygen atom. As a matter of fact, the mass of the hydrogen atom (the lightest of all atoms) is, to use the figure now accepted as most nearly accurate, 1.67343×10^{-24} grams or, in non-exponential form 0.00000000000000000000-000167343 grams.

From Avogadro's number, it is also possible to calculate the diameter of atoms if one assumes that they are spherical in shape and that, in liquids and solids, they are packed together in virtual contact. It then turns out that the diameter of atoms is approximately 10^{-8} centimeters. In ordinary terms, this means that 250,000,000 atoms placed side by side would make a line an inch long.

With atoms so small and so light, it is no wonder that matter seems continuous to our senses and that men like Democritus, who postulated atoms on purely philosophic grounds, found it so difficult to persuade others of the value of their suggestion.

But even the determinations of the mass and size of the atom rest on indirect evidence. In ordinary life, reality is judged by the direct evidence of the senses—especially that of vision. "Seeing is believing," goes the old bromide.

It is, of course, quite possible to argue that seeing is not necessarily believing; that hallucinations and optical illusions are possible; and that it is not always easy to interpret what one sees (as when one "sees" that the earth is flat). It follows, then, that careful and logical reasoning based on a large accumulation of accurate, but indirect, data can be a more reliable guide to useful conclusions than the senses may be.

Nevertheless, human prejudices being what they are (even among scientists), it is rather exciting to know that atoms have been made visible, at least after a fashion. This came about through the invention by the German-American physicist Erwin Wilhelm Mueller (1911–   ) of specialized forms of powerful microscopes.

The first of these, devised in 1936, was the *field-emission microscope*. This begins with a very fine needle-tip enclosed in a high vacuum. Under an intense electric field, such a needle can be made to shoot out very tiny particles.* If only these particles would travel in perfectly straight, undeviating lines to a screen enclosing the vacuum tube, they would produce a pattern that would depict the actual atomic makeup of the needle-tip. Unfortunately, in even the best vacuums there are gas molecules here and there. The flying particles that strike these molecules are diverted. The result is a fuzzy, out-of-focus picture.

In the 1950's, Mueller made use of heavier particles. He introduced small quantities of helium atoms. When any of these struck the needle-tip, they were modified by the electric field into helium ions (see page 27) which then raced away from the needle-tip in a straight line.

The heavy helium ions are not easily diverted even by collisions with gas molecules, and a much sharper picture is obtained in such a *field-ion microscope*. The atoms in the needle-tip are then pictured as round dots arranged in orderly and well-packed fashion. This device is applicable only to a limited number of high-melting metals, but it has the effect of making atoms visible and therefore "real." Several photographs of the atom patterns revealed in this fashion have already become scientific classics.

* These particles are called electrons and are even smaller than atoms. They will be discussed in detail throughout this book.

2

Ions and Radiation

With 103 different elements now known and, therefore, 103 different kinds of atoms, there is good reason to feel uncomfortable. The periodic table imposes an order upon them, to be sure, but why should that particular order exist?

Why are there so many elements? Why should slight differences in mass between two sets of atoms make so much difference? For instance, argon has an atomic weight of 39.9 and potassium one of 39.1, and yet that small difference makes the first a very inert gas and the second a very active metal.

.To obtain an understanding of atomic properties, one might attempt to delve within the atom. One might wonder whether the atoms might not themselves have a structure and whether the atom might not best be understood in terms of this structure.

Something of this sort occurred in 1816, quite early in the game, to an English physician, William Prout (1785–1850). At the time, the atomic theory was very new and the only atomic weights known were a few that had been determined (not very accurately) by Dalton. These atomic weights, based on a hydrogen = 1 standard, were all integers.

To Prout, this seemed more than one could expect of coincidence. If all the atoms had masses that were integral multiples of the mass of the hydrogen atom, then was it not reasonable to suppose that the more massive atoms were made up of hydrogen atoms? If oxygen had an atomic weight of 16, for instance, might not this be because it was made up of 16 hydrogen atoms tightly mashed together?

Prout published this suggestion anonymously, but his authorship became known and his explanation has been called *Prout's hypothesis* ever since.

For a century afterward, numerous chemists made accurate atomic weight determinations for the purpose (in part, at least) of checking on whether or not they were all integral multiples of the atomic weight of hydrogen. They proved not to be. As stated earlier (see page 12) the oxygen atom was not 16 times as massive as the hydrogen atom, judging by atomic weight determinations, but 15.88 times. The atomic weight of nickel is 58.24 times that of hydrogen, and so on.

Over and over again, Prout's hypothesis was disproved and yet, with the opening of the twentieth century, chemists were still uneasy about it. About half the elements had atomic weights that were quite close to integral values. This was still asking a great deal of coincidence. Surely there had to be significance in this fact.

There was, of course, and that significance was discovered in very roundabout fashion through a line of investigation that began with electricity.*

It was in 1807 and 1808 that Humphry Davy had produced a series of elements (sodium, potassium, calcium, magnesium, strontium, and barium) by passing an electric current through molten compounds that contained atoms of these elements in their molecules. The work was carried on with greater detail by the English chemist Michael Faraday (1791–1867), who in his youth had been Davy's assistant and protégé.

Imagine two metal rods connected to a battery, one to the positive pole, the other to the negative pole. These rods are *electrodes* (from Greek words meaning "the path of the electricity"). Faraday called the one attached to the positive pole the *anode* ("upper path") and the one attached to the negative pole the *cathode* ("lower path"). (Electricity at the time was assumed to flow from the positive pole to the negative pole, like water flowing from an upper level to a lower one.)

If the two electrodes are brought together and allowed to touch, electricity will flow through them. However, if they are

* Electricity makes up the subject matter of the second half of Volume II

separated by an air gap, the circuit is broken and electricity will not flow. If the electrodes are not in contact but are both immersed in the same container of liquid, electricity may or may not flow, depending on the nature of the liquid. Immersed in a dilute solution of sulfuric acid or of sodium chloride, current will flow; immersed in a dilute solution of sugar or in distilled water, current will not flow. The former liquids are conductors of electricity, the latter are nonconductors. Faraday called the liquid conductors *electrolytes*, and the liquid nonconductors, *nonelectrolytes*.

The passage of an electric current through an electrolyte induces chemical changes. Often these changes consist of the decomposition of some of the molecules contained in the solution and in the production of elements (*electrolysis*), as in the case of the metals produced by Davy from their compounds.

The elements, when produced, appear at the electrodes. If they are gases, they bubble off. If they are metals, they remain clinging to the electrode (*electroplating*.)

Elements can appear at either electrode. If electricity passes through water containing a bit of sulfuric acid, hydrogen appears at the cathode and oxygen at the anode. If an electric current passes through molten salt (sodium chloride), metallic sodium appears at the cathode, gaseous chlorine at the anode.

Faraday did not allow himself to speculate too freely about the exact manner in which an element was transported through the body of the solution to one electrode or the other. One might think of drifting atoms, but Faraday was rather lukewarm on the atomic theory (still new at the time of his experiments) and he preferred not to commit himself. He spoke simply of *ions* (from a Greek word meaning "wanderer") passing through the solution, and said nothing about their nature.

Some ions, like those which ended as sodium or hydrogen, are attracted to the cathode; they are *cations* (pronounced in three syllables). Others, like those which end as chlorine or oxygen, are attracted to the anode and are *anions* (again three syllables).

Faraday carefully measured the mass of element produced by the action of the electric current and, in 1832 and 1833, proposed what have since become known as *Faraday's laws of electrolysis*.

The first law of electrolysis states: The mass of element formed by electrolysis is proportional to the quantity of electric current passing through an electrolyte. The unit of quantity of electricity in the meter-kilogram-second (mks) system is the *cou-*

lomb (see page II–164), and one coulomb of electricity will form 0.001118 grams of metallic silver when passed through a solution of a silver compound. By Faraday's first law two coulombs of electricity would produce twice that mass of silver and, in general, x coulombs will produce 0.001118x grams of silver.

A gram-atomic weight of silver is equal to 107.87 grams. How many coulombs would be required to deposit that many grams? It is only necessary to set 0.001118x 107.87 and solve for x, which turns out to be equal to about 96,500 coulombs. For this reason, the quantity of electricity represented by 96,500 coulombs is set equal to one *faraday*. The faraday may be defined as that quantity of electricity which will liberate one gram-atomic weight of metallic silver from a silver compound.

To understand Faraday's second law of electrolysis, it is first necessary to grasp the meaning of *equivalent weight*.

One gram-atomic weight of chlorine gas (35.5 grams) will combine with one gram-atomic weight of hydrogen (1 gram) to form hydrogen chloride (HCl). The molecule is composed of one atom of each element, and since a gram-atomic weight of hydrogen and a gram-atomic weight of chlorine contain the same number of atoms of those elements, the two quantities of gas match up neatly. (The fact that in the case of both hydrogen and chlorine the atoms happen to be distributed in the form of two-atom molecules does not alter the case.) By the same reasoning, one gram-atomic weight of chlorine will combine with one gram-atomic weight of sodium (23 grams) to form sodium chloride (NaCl.)

However, one gram-atomic weight of chlorine will combine with only half a gram-atomic weight of calcium to form calcium chloride ($CaCl_2$) because every calcium atom takes up two chlorine atoms; consequently only half as many calcium atoms as chlorine atoms are needed for the reaction. The gram-atomic weight of calcium is 40 grams, and half a gram-atomic weight is 20 grams. This means that 20 grams represents the equivalent weight of calcium: the weight that is equivalent, that is, to a gram-atomic weight of chlorine or of hydrogen or of sodium in forming compounds. (It is usually the gram-atomic weight of hydrogen which is taken as the standard.)

In the same way one gram-atomic weight of chlorine will combine with half a gram-atomic weight of magnesium to form magnesium chloride ($MgCl_2$) and with a third of a gram-atomic weight of aluminum to form aluminum chloride ($AlCl_3$). The equivalent weight of magnesium is its gram-atomic weight (24 grams) divided by 2, or 12 grams, while that of aluminum is its gram-atomic weight (27 grams) divided by 3, or 9 grams.

Now we can return to Faraday's second law of electrolysis, which can be stated most simply, as follows: One faraday of electricity will form an equivalent weight of an element when passing through a compound of that element.

If a faraday of electricity will form 108 grams of silver, it will also form 23 grams of sodium, 35.5 grams of chlorine, or 1 gram of hydrogen (in each case equal to the gram-atomic weight). It will form 20 grams of calcium or 12 grams of magnesium (in each case equal to half the gram-atomic weight). It will form 9 grams of aluminum (equal to a third the gram-atomic weight).

Particles of Electricity

Faced with these laws of electrolysis, it is extremely tempting to begin wondering whether electricity might not be particulate in nature. Just as matter consists of indivisible units (atoms), so might electricity.

Let us assume this is so, and let us further assume that such units come in two varieties. There is a positive unit that is attracted to the negatively-charged cathode (opposite electric charges attract, see page II–159). It is such a positive unit that can carry atoms of hydrogen and sodium in the direction of the cathode. Similarly there would be a negative unit that is attracted to the positively-charged anode and that can carry atoms of oxygen and chlorine with it. The two units can be symbolized as + and −.

If we imagine a hydrogen atom being transported toward the cathode by a positive electrical unit, we can symbolize the hydrogen atom in transit as H^+. Using Faraday's term, we can call it a *hydrogen ion*. Similarly, we can have a sodium ion (Na^+) or a potassium ion (K^+). All three are examples of *positive ions* (or cations).

A faraday of electricity can be viewed as containing Avogadro's number of electrical units. Allowing one unit per atom, a faraday of electricity would transport Avogadro's number of hydrogen atoms to the cathode. In other words, a faraday of electricity would produce a gram-atomic weight of hydrogen at the electrode. It would also, by similar reasoning, produce a gram-atomic weight of sodium atoms or potassium atoms or silver atoms.

Since a faraday of electricity has never, under any conditions, been found to produce more than a gram-atomic weight of any element, it seems reasonable to conclude that the electric unit we are dealing with is very likely the smallest unit possible—that it is an indivisible unit and that one unit can transport no more than one atom.

Chlorine atoms are transported to the positive electrode, or anode, and therefore must be transported by a negative electric unit. We can symbolize the chlorine atom in transport as Cl^- and call that the *chloride ion*.* Since a faraday of electricity produces exactly one gram-atomic weight of chlorine, we must conclude that the negative unit is exactly equal in size to the positive unit

What of calcium? A faraday of electricity will produce only half a gram-atomic weight of that element. This is most easily explained by assuming that the atom, on its travels toward the cathode, must be transported by two positive units. In that case the supply of units in a faraday of electricity will only transport half the number of atoms of calcium one would expect if one were dealing, say, with sodium. We can write the *calcium ion* as Ca^{++}, therefore. By similar reasoning, we can write the *magnesium ion* as Mg^{++}, the *barium ion* as Ba^{++}, the *aluminum ion* as Al^{+++}, the *oxide ion* as O^{--}, and so on.

The first to maintain, in complete and logical detail, that Faraday's ions were actually atoms carrying a positive or negative electric charge, was the Swedish chemist Svanté August Arrhenius (1859–1927). These views, presented first in 1887, were based not only on Faraday's work but on other chemical evidence as well.

According to Arrhenius, when an electric current passed through molten sodium chloride, the molecule (NaCl) broke up, or dissociated†, not into atoms, but into charged ions, Na^+ and Cl^-, the sodium ions then drifting toward the cathode and the chloride ions toward the anode. (This is the *theory of ionic dissociation*.) At cathode and anode, the ions are discharged and the uncharged atoms are produced; metallic sodium at the cathode, gaseous chlorine at the anode.

The charged atom, Arrhenius maintained (correctly, as it turned out), did not necessarily have properties in any way resembling those of the uncharged atom. Sodium atoms, for

---

* It is called "chloride ion" rather than "chlorine ion" for reasons involving chemical nomenclature. These are more fittingly discussed in a book on chemistry. For our purposes here, we can take chemical names as we find them.

† It has since turned out that Arrhenius was wrong is assuming that this dissociation into ions took place only under the influence of the electric current. The atoms in sodium chloride exist in ionic form at all times. However, Arrhenius, like Dalton, was right enough for his time.

instance, would react violently with water, but sodium ions, much milder in character, would not. Chlorine atoms would form chlorine molecules and bubble out of solution; chloride ions would not.

It followed further from Arrhenius' analysis that groups of atoms, as well as individual atoms, might carry an electric charge. Thus, ammonium chloride (NH_4Cl) will dissociate to form NH_4^+ and Cl^-, the former being the *ammonium ion*. Again, sodium nitrate ($NaNO_3$) will break up into Na^+ and NO_3^-, the latter being the *nitrate ion*. Other such *compound ions* (those made up of more than one atom) are the *hydroxyl ion* (OH^-), the *sulfate ion* (SO_4^{--}), the *carbonate ion* (CO_3^{--}) and the *phosphate ion* (PO_4^{---}).

So much in the air was this notion of an indivisible unit of electricity that the Irish physicist George Johnstone Stoney (1826–1911) had even given it a name in a paper published in 1881. He called it an *electron*.

Despite the logic of Arrhenius' views (especially as viewed from hindsight), his theory of ionic dissociation was met with great reserve. The notion of an atom as a featureless, structureless, indivisible object dated back to Democritus and had become a firm part of scientific thinking. The thought of such atoms carrying indivisible units of electric charge ("atoms of electricity" so to speak) was hard to take without heavy evidence on its side.

Such evidence was not obtained in completely acceptable form for a decade after Arrhenius, but it was on its way in Arrhenius' time and even before.

The chief difficulty in detecting particles of electricity under ordinary conditions was that even supposing they existed, they would be lost among the ordinary particles of matter in the path of the electric current.

What was clearly needed was the passage (if possible) of an electric current through a good vacuum. Then the particles of electricity (if any) might show up unmasked. The first to actually force a current of electricity through a vacuum was Faraday himself, in 1838. However, the best vacuum he could obtain was not a very good one, and his observations therefore lacked significance.

In 1854, a German glassblower, Heinrich Geissler (1814–1879), devised a better method for producing vacuums than any hitherto obtained. He manufactured *Geissler tubes* containing these good vacuums. The German physicist Julius Plücker (1801–1868) made use of such Geissler tubes into which two electrodes had been sealed.

Plücker forced electricity to cross the vacuum from one electrode to the other and noted that a greenish luminescence coated the cathode when the current flowed. This greenish luminescence seemed precisely the same whatever the metal out of which the cathode was constructed and whatever the nature of the wisps of gas that still remained in the tube after it had been evacuated. Whatever that luminescence might be, then, it was a property of electricity and not of ordinary matter.

Plücker also showed that the luminescence shifted its position when a magnet was brought near. One pole of the magnet shifted it in one direction; the other pole, in the opposite direction. This also seemed to brand the luminescence an electrical phenomenon, since electricity and magnetism are very closely allied (see page II–237).

It soon became obvious that the phenomenon was not merely confined to the near neighborhood of the cathode but that something was traveling all across the space from the cathode to the anode. What's more, this something traveled in straight lines. If the anode were placed to one side, whatever it was that was traveling missed the anode and went on to strike the glass of the tube, creating a spot of green luminescence where it struck.

Two investigators, the German physicist Johann Wilhelm Hittorf (1824–1914) and the English physicist William Crookes (1832–1919), working independently, showed that if in such a tube an object was enclosed in the path of the traveling entity, that object cast a shadow against the luminescence on the glass. Hittorf published his results first—in 1869.

It was clear, then, that physicists were faced with a kind of radiation that traveled in straight lines and cast sharp shadows. The German physicist Eugen Goldstein (1850–1930), committing himself no further than this and taking note of the appar-

Crookes tube

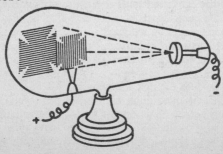

ent origin of the radiation, called it *cathode rays* in 1876. This name was generally adopted.

A controversy then arose as to the nature of the cathode rays. The fact that the rays traveled in straight lines and seemed unaffected by gravity made it appear likely that they were a wave form after the fashion of light. The great argument against this was that the cathode rays were deflected by a magnet, whereas light rays (or any form of radiation resembling light) were not.

The alternative suggestion was that the cathode rays were electrically charged particles, the "atoms of electricity" in fact. They would then naturally be affected by a magnet, and their lack of response to gravitation would be explained by their small mass and rapid motion. The response would be there but would be too small to detect.

The Radiation Spectrum

The controversy over the nature of the cathode rays divided itself almost on national lines, with many German physicists upholding the wave interpretation and many English physicists maintaining the charged-particle suggestion.

This was a natural division, perhaps, for it was in Germany that indisputably new wave forms were discovered in the final decades of the nineteenth century—although the first such discovery was inspired by the theory of an Englishman, James Clerk Maxwell.

Maxwell's analysis of electrical and magnetic phenomena had showed that the two must be so closely and indissolubly related that one could properly speak only of electromagnetism. He went on to show, furthermore, that an oscillating electric charge ought to produce a wave-form type of *electromagnetic radiation* that would travel at the speed of light. It seemed almost inevitable, therefore, that light itself must be an electromagnetic radiation— otherwise the coincidence of its velocity being equal to that of such radiation would be too great for acceptance.

But if Maxwell was correct, there was no reason why man could not deliberately produce an electromagnetic radiation by oscillating an electric current. It could not be oscillated fast enough to produce the tiny wavelengths of light (that would have required about a quadrillion oscillations per second), but Maxwell's theory set no limit on the period of oscillation. A comparatively slow oscillation of, say, a thousand times per second would produce a thousand waves of electromagnetic radiation per second. Since the wave train would travel 300,000 kilometers per second, each individual wave would be 300 kilometers long (vastly longer than the wavelengths of light), but those waves would nevertheless exist.

The attempt to form long wavelength radiation was made in 1887 by a German physicist, Heinrich Rudolf Hertz (1857–1894). He set up an electric circuit that would produce a spark across a small air gap under conditions that would bring about an electrical oscillation of the sort that Maxwell said would produce electromagnetic radiation. To detect the radiation, if any was produced, Hertz used a simple rectangle of wire broken by a small air gap. The electromagnetic radiation crossing the wire receiver would cause an electric current to flow and produce a spark across the air gap.

Such a spark was found and Hertz knew he had detected the electromagnetic radiation predicted by Maxwell—a result that served as strong evidence in favor of the validity of Maxwell's theory. At first, the radiation discovered by Hertz was called "Hertzian waves." However, the noncommittal title of *radio waves* ("waves that radiate") is now usually used.

The discovery of radio waves gave physicists their first notion of the truly broad extent of the electromagnetic spectrum. The wavelength range of visible light is from 380 to 760 millimicrons, representing a single octave of radiation. (A millimicron is equal to a billionth of a meter, and an octave represents a range over which the wavelength doubles.)

It was not until 1800 that this spectrum was broadened beyond the visible. In that year, the German-English astronomer William Herschel (1738–1822) was measuring the effect of the solar spectrum upon a thermometer. He discovered that the temperature-raising effect of the spectrum was most marked at a point somewhat beyond the red, where the eye could see nothing. Herschel correctly concluded that light was present there—light which was incapable of affecting the retina of the eye.

At first, because of the efficient manner in which the glass and mercury of the thermometer absorbed this invisible light, it was referred to as "heat rays." Later, the more noncommittal term, *infrared radiation* ("below the red") was used. With the establishment of the wave theory of light (see page II–66) it was understood that infrared radiation was of longer wavelength than visible light.

Nowadays, the range of infrared radiation is taken as extending from the 760 millimicron limit of visible light to a rather arbitrary upper limit placed at 3,000,000 millimicrons. In expressing the wavelength of infrared radiation, the more convenient unit of length, the micron (equal to 1000 millimicrons), may be used. The range of infrared radiation can then be said to extend from 0.76 microns to 3000 microns, a stretch of about 12 octaves.

Beyond the farthest infrared radiation lie the radio waves. The radiation found in the wavelength region immediately adjacent to the infrared has come in recent years to be known as *microwaves* ("small waves"—small for radio waves, that is). The microwave region extends from 3000 to 300,000 microns. Again we can shift units to millimeters (one millimeter is equal to a thousand microns) and say that the range is from 3 to 300 millimeters, or about 6½ octaves.

Beyond the microwaves are the radio waves, proper. For them there is no definite upper limit. Radio waves of longer and longer wavelength can easily be produced until they become too low in energy to detect by presently available means. (The longer the wavelength of electromagnetic radiation, the lower its energy content, as the quantum theory makes plain, see page II–130). Radio waves up to 30,000,000 millimeters in wavelength have been used in technology, so that we can say that useful radio waves extend over a range of from 300 to 30,000,000 millimeters (0.3 to 30,000 meters), or 16½ octaves.

The electromagnetic spectrum also extends out beyond the violet end of the visible light region. This was first discovered in 1801 by the German physicist Johann Wilhelm Ritter (1776–1810), who was studying the action of light upon silver nitrate. Silver nitrate, white in color, breaks down in the presence of light, liberating black particles of metallic silver and turning visibly gray in consequence. This effect is more marked where shortwave light impinges on the silver nitrate (which is not surprising to us now since shortwave light is known to be the more energetic, so that it naturally initiates an energy-consuming chemical reaction more readily). Ritter discovered that the effect on silver nitrate was even more pronounced when the compound was placed beyond the violet end of the solar spectrum where nothing at all could be seen.

Like Herschel, Ritter concluded that invisible light was present. Because of its effect on silver nitrate it was at first referred to as "chemical rays." This soon gave way, however, to *ultraviolet radiation* ("beyond the violet"), and it came to be understood that such radiation was shorter in wavelength than visible light was.

Nowadays, ultraviolet radiation is taken as covering a range of from 360 millimicrons (the boundary of the visible violet) down to an arbitrary limit of 1 millimicron, a little over eight octaves. Thus, as the 1890's opened, the overall stretch of the electromagnetic spectrum, from ultraviolet radiation to radio waves, represented an extreme range of some 44 octaves, of which only one was visible light.

Even so, the electromagnetic spectrum had not been completely filled in. The next step was taken by the German physicist Wilhelm Konrad Röntgen (1845–1923). He was interested in cathode rays and, in particular, in the luminescence to which they gave rise when they impinged on certain chemicals.

In order to observe the faint luminescence, he darkened the room and enclosed the cathode-ray tube in thin black cardboard. On November 5, 1895, he set the enclosed cathode-ray tube into action, and a flash of light that did not come from the tube caught his eye. He looked up and quite a distance from the tube he noted a sheet of paper that had been coated with barium platinocyanide (a compound that glows under the impact of energetic radiation) shining away. Röntgen would not have been surprised to see it glow if cathode rays were striking it, but the cathode rays were completely shielded off.

Röntgen turned off the tube; the coated paper darkened. He turned it on again; it glowed. He walked into the next room with the coated paper, closed the door, and pulled down the blinds. The paper continued to glow while the tube was in operation.

It seemed to Röntgen that some sort of radiation was emerging from the cathode-ray tube, a radiation produced by the impact of the cathode rays on the solid material with which it collided. The kinetic energy lost by the cathode rays as they were stopped was converted, apparently, into this new form of radiation; a radiation so energetic that it could pass through considerable thicknesses of paper and even through thin layers of metal. Röntgen published his first report on the subject on December 28, 1895.

The radiation is sometimes known as "Röntgen rays" after the discoverer, but Röntgen himself honored its unknown nature by using the mathematical symbol of the unknown and named the radiation *X rays*. The name has clung firmly even though the nature of the radiation is no longer mysterious.

I have mentioned the experiment in some detail because the discovery of X rays is usually taken as initiating the "Second Scien tific Revolution" (the first having been initiated by the experiments of Galileo—see page I-9).

In a way, this might be considered as over-dramatic, for Röntgen's experiment did not really represent a sharp break with previous work. It came about in connection with the cathode ray problem, which was occupying many physicists of the time. The new radiation had been observed by Crookes and Hertz even before Röntgen's announcement (although they had not grasped the significance of what they were observing), so that the discovery was inevitable. If Röntgen had not made it, someone else would have done so, perhaps within weeks. Moreover, the existence of X rays was implicit in Maxwell's theory of electromagnetic radiation, and the important discovery in that connection, the one that validated the theory, was that of radio waves, eight years earlier.

Nevertheless, after all this has been allowed for, the X rays caught the fancy of both the scientists and the lay public with unprecedented intensity. Their ability to penetrate matter was fascinating. On January 23, 1896, in a public lecture, Röntgen took an X ray photograph of the hand of the German biologist Rudolf Albert von Kölliker (1817–1905), who volunteered for the purpose. The bones showed up beautifully, for they stopped the X rays where flesh and blood did not. The photographic film behind the soft tissues was fogged by the X rays that reached it, while the portion of the film behind the bone was not. The bones showed up, therefore, as white against gray.

The usefulness to medicine was obvious and X rays were at once applied there and in dentistry. (The dangers of X rays as cancer-producing agents were not understood for a number of years.) A storm of experimentation involving X rays followed, and as a result, discoveries were made which resulted in so quick and rapid an improvement of man's understanding of the universe that it seemed a scientific revolution indeed.

CHAPTER **3**

The Electron

The Discovery of the Electron

Since of the new types of radiation, the radio waves were known to be wave forms, and the X rays were strongly suspected of being wave forms (final proof that they were was obtained in 1912, see page 63) it seemed all the more natural to continue to suspect that cathode rays were also wave forms.

For one thing, Hertz showed in 1892 that cathode rays could actually penetrate thin sheets of metal. This seemed quite an unlikely property for particles to possess, whereas a few years later the discovery of X rays made it quite clear that wave forms could possess this property. The German physicist Philipp Lenard (1862–1947), Hertz's assistant, even set up a cathode-ray tube containing a thin metal "window." Cathode rays striking that window passed through and emerged into open air. (Such emerging cathode rays were for a time called "Lenard rays.")

If cathode rays were electrically charged particles, they should be affected not only by a magnetic field but also by an electrostatic field. Hertz passed a beam of cathode rays between two parallel plates, one carrying a positive electric charge and another, a negative one. He detected no deviation in the cathode ray stream and concluded that cathode rays were waves.

That, however, marked the peak of the wave theory. Another experimenter was on the scene, a member of the English group of physicists, Joseph John Thomson (1856–1940). It seemed to him that the experiment involving the electrostatic field would not work unless the cathode rays were passing through a particularly good vacuum. Otherwise the thin wisps of gas present would, by Thomson's reasoning, act to reduce the effect of the electrostatic field upon the cathode rays. In 1897, he therefore repeated Hertz's experiment (Hertz having prematurely died three years earlier), using a cathode-ray tube with a particularly good vacuum. A deflection in the path of the cathode rays was detected.

This observation was the last straw. With cathode rays deflected by both a magnetic field and an electrostatic field, the evidence in favor of particles was too strong to be withstood. From the direction of the deflection, it could be seen that the particles carried a negative charge.

It seemed clear that these cathode ray particles must represent units of electricity, perhaps the indivisible negative unit (see page 29) which some nineteenth century physicists had been postulating. The particles were therefore given Stoney's name of "electron," and it is because of Thomson's crucial experiment that he is usually said to have "discovered the electron" in 1897.

But Thomson did more than merely discover the electron. He went on to determine one of its overwhelmingly important properties.

When an electron passes through a magnetic field, it is deflected by that field and departs from its otherwise straight-line course to take up a curved path. (This is analogous to the fashion in which the moon, when exposed to the gravitational field of the earth, departs from what would otherwise be a straight-line course to take up a curved path.)

The deflection of the electron is the result of the magnetic force exerted upon it. The amount of this force is proportional, first to the strength of the magnetic field (H), then to the size of the electric charge on the electron (e), and finally to the velocity of the electron (v)—for it is the velocity that determines how many magnetic lines of force will be cut by the moving electron. (A stationary electron, or one traveling parallel to those lines of force, would not be affected by the magnetic field.) The force producing the deflection is therefore equal to Hev.

A centrifugal effect must be exhibited by an electron traveling in a curved path. This effect is equal to mv^2/r, where m is the mass of the electron, v its velocity, and r the radius of the curved path it is following.

The electron, in following a particular curved path, must have the magnetic force exactly balanced by the centrifugal effect. If this were not so, it would travel either a tighter curve or a looser one, finding a curve in which the two effects would balance. For the curved path actually followed, we can therefore say that:

$$Hev = \frac{mv^2}{r} \qquad \text{(Equation 3–1)}$$

This can be rearranged and simplified to:

$$\frac{e}{m} = \frac{v}{Hr} \qquad \text{(Equation 3–2)}$$

The strength of the magnetic field is known, and the curvature radius of the beam of cathode ray particles can easily be determined by the shift in the position of the luminescent spot on the wall of the cathode-ray tube. Now if one could only determine the value of v (the velocity of the particles), it would at once be possible to determine the value of e/m (the ratio of the charge of the electron to its mass).

Thomson found the velocity by causing the cathode rays to be under the influence of both an electrostatic field and a magnetic field, but under such conditions that the two deflections were in opposite directions and just balanced. The deflection by the electrostatic field depended upon its strength (F) and upon the charge of the electron (e). It did not depend on the velocity of the electron, for there is attraction between opposite electric charges even if they are stationary relative to each other.

Consequently, when the magnetic and electrostatic fields are adjusted in strength so that the effects on the electrons cancel out:

$$Hev = Fe \qquad \text{(Equation 3–3)}$$

or:

$$v = \frac{F}{H}. \qquad \text{(Equation 3–4)}$$

Since the strength of both fields can be measured easily, v can be determined and turns out to be about 30,000 kilometers per second, about one-tenth the velocity of light. This was by far the largest velocity ever measured for material objects up to that time and immediately explained why cathode rays had seemed unaffected by a gravitational field. At that enormous velocity, particles passed from end to end of the cathode-ray tube long before they could show a measurable response to the earth's gravitational field.

With the value of v known, Equation 3–2 makes it at once possible to determine e/m for the electron, and Thomson was amazed to discover that this ratio came out to have a value, far greater than that for any ion (which are also charged particles).

Consider the ions H^+, Na^+ and K^+. All three carry a charge of equal magnitude since a faraday of electricity suffices to produce a gram-atomic weight of each. However, the mass of the potassium ion is 39 times that of the hydrogen atom and the mass of the sodium atom is 23 times that of the hydrogen atom. If e is fixed, then the ratio e/m rises as m decreases. Thus the ratio e/m for H^+ must be 23 times as great as that for Na^+, and 39 times as great as that for K^+.

Indeed, since the hydrogen ion is the least massive ion known, the e/m ratio for it may well be higher than for any other ion that can possibly exist. And yet the e/m ratio for the electron is (using the value now accepted) 1836 times as great as that of the hydrogen ion.

No quantity of electric charge smaller than that on the hydrogen ion had ever been observed, and it seemed reasonable to suppose that the electron carried this smallest-observed charge. If that is so, if e is equal in the case of the electron and the hydrogen ion, and e/m is 1836 times greater in the first case than in the second, it must follow that the difference is to be found in the mass. The mass of the electron must be only 1/1836 that of the hydrogen ion.

Since the mass of the hydrogen atom is known and the mass of the hydrogen ion is only very slightly less, it is easy to calculate the mass of the electron. The best modern determination is 9.1091 $\times 10^{-28}$ grams, or 0.00000000000000000000000000091091 grams.

In one bound, the atoms, which from the time of Democritus on had been assumed to be the smallest particle of matter, were suddenly rendered giants. Here was something much smaller than even the smallest atom; something so small indeed that it could easily be visualized as worming its way through the interstices among the atoms of ordinary matter. That seemed one reasonable explanation for the fact that cathode rays made up of particles could penetrate thin sheets of metal. It also explained why electric currents could be made to flow through copper wires.

Thomson, therefore, had not only discovered the electron, he had also discovered the first of the *subatomic particles*, and opened a new realm of smallness beyond the atom.

The Charge of the Electron

Knowledge of the exact mass of the electron did not automatically provide physicists with an estimate of the exact size of the charge upon the electron. One could only say, at first, that the charge on the electron was exactly equal to the charge on the chloride ion, for instance, or exactly equal (but opposite in sign) to the charge on the hydrogen ion. But then, the exact size of the charge on any ion was not known through the first decade of the twentieth century.

The experiments that determined the size of the electric charge on the electron were conducted by the American physicist Robert Andrews Millikan (1868–1953) in 1911.

Millikan made use of two horizontal plates, separated by about 1.6 centimeters, in a closed vessel containing air at low pressure. The upper plate had a number of fine holes in it and was connected to a battery that could place a positive charge upon it. Millikan sprayed fine drops of a nonvolatile oil into the closed vessel above the plates. Occasionally, one droplet would pass through one of the holes in the upper plate and would appear in the space between the plates. There it could be viewed through a magnifying lens because it was made to gleam like a star through its reflection of a powerful beam of light entering from one side.

Left to itself, the droplet of oil would fall slowly, under the influence of gravity. The rate of this fall in response to gravity, against the resistance of air (which is considerable for so small and light an object as an oil droplet), depends on the mass of the droplet. Making use of an equation first developed by the British physicist George Gabriel Stokes (1819–1903), Millikan could determine the mass of the oil droplets.

Millikan then exposed the container to the action of X rays. This produced ions in the atmosphere within (see page 110). Occasionally, one of these ions attached itself to the droplet. If it were a positive ion, the droplet, with a positive charge suddenly added, would be repelled by the positively-charged plate above, and would rush downward at a rate greater than could be accounted for by the action of gravity alone. If the ion were negative, the droplet would be attracted to the positively-charged plate and might even begin to rise in defiance of gravity.

The change in velocity of the droplet would depend on the intensity of the electric field (which Millikan knew) and the charge on the droplet, which he could now calculate.

Millikan found that the charge on the droplet varied according to the nature of the ion that was adsorbed and on the number of ions that were adsorbed. All the charges were, however, multiples of some minimum unit, and this minimum unit could reasonably be taken as the smallest possible charge on an ion and, therefore, equal to the charge on the electron. Millikan's final determination of this minimum charge was quite close to the value now accepted, which is 4.80298×10^{-10} electrostatic units ("esu," see page II–164), or 0.000000000480298 esu.

As far as we know now, this charge of 4.80298×10^{-10} esu is the only size in which electric charge comes, though it may come in two varieties of that size, positive and negative. Suppose we consider this charge unit as 1, for simplicity's sake. In that case, all objects can be placed in one of three classes:

(1) Objects with a net electric charge of 0. This would include ordinary atoms and molecules.

(2) Objects with a net charge of -1, or some multiple of that. Examples are some negative ions, and, of course, the electron.

(3) Objects with a net charge of $+1$ or some multiple of that. Examples of that are some positive ions.

No one has yet discovered an object with a charge of $+0.5$ or -1.3 or, in fact, with a charge that deviates from an integral value by even the slightest. Such objects may yet be discovered in the future, but the prospects for such an eventuality seem quite small at the moment.

Electronics

It was the existence of electrons, and of subatomic particles generally, that was to bring a new degree of order into the table of elements. Before proceeding in that direction, however, let us consider some of the changes in technology that arose out of the use of streams of electrons in a vacuum. (The study of the behavior of such free electrons and of the techniques for controlling and manipulating them is called *electronics*.)

The flow of electrons across a vacuum was observed under

interesting circumstances in 1883 by the American inventor Thomas Alva Edison (1847–1931). Four years earlier he had devised a practical electric light, and he was still laboring to improve it. The light, at that time, consisted of a carbon filament enclosed in an evacuated bulb. (The vacuum was necessary in order to keep the carbon filament, raised to white-hot temperatures by the current passing through it, from burning to nothing in a flash—as it would if air were present.)

Edison observed on the interior surface of the bulb, a blackening which presumably resulted because some of the carbon vaporized from the hot filament surface and settled on the glass. This weakened the filament and reduced the transparency of the glass, so Edison sought to counter the effect. One of his efforts to do so consisted of sealing a small strip of metal into the bulb near the filament, hoping perhaps that the metal would blacken rather than the glass.

This did not happen, but Edison noticed something else. When he attached this piece of metal (called a *plate* by later workers) to the positive pole of a battery, so that it took on a positive charge with respect to the filament, a current flowed even though there was a gap in the circuit between the filament and the plate. If the plate was given a negative charge, this did not happen. Edison described this phenomenon (the *Edison effect*) and then, since he had no immediate use for the matter, laid it aside.

The Edison effect was no mystery once the cathode rays were understood. The heated filament had a tendency to give off electrons; they "boiled off," so to speak. Ordinarily, this would result in no more than a thin cloud of electrons surrounding the filament.

If, however, a positively-charged plate was placed in the neighborhood, the electrons would be attracted to it. A stream of electrons would pass continuously from the heated filament to the plate, and this is equivalent to a completed electric circuit. If the plate is negatively charged, the electron cloud is repelled, and the circuit is not completed; there is no flow of electricity.

An English electrical engineer, John Ambrose Fleming (1849–1945), who had served as consultant to Edison in the 1880's, remembered the Edison effect twenty years later, in 1904. Suppose the plate was attached to an alternating-current circuit (see page II–221). When the current flowed in one direction, the plate would receive a positive charge; when it flowed in the other, it would receive a negative charge. The nature of the charge would shift some sixty times a second in sixty-cycle alternating current. However, only when the plate was positively charged would the circuit really be complete.

Half the time, then, when the current was flowing in one direction, it would actually flow. The other half the time, when it would ordinarily be expected to flow in the other direction, it would not flow at all, for the circuit would be broken.

The Edison effect made it possible for the circuit to be opened and closed in exact time with the alternation of the current. What would have been an alternating current without the filament-plate combination in the circuit, becomes a direct current with it. The current might flow only intermittently and with fluctuating intensity, to be sure, but it would always flow (when it did flow) in the same direction. The filament-plate combination acted as a *rectifier*.

Fleming called the device a "valve" because it opened and shut the gate to the flow of electricity as an ordinary valve might do for a flow of water. In the United States, the far less significant name of *vacuum tube* has come into use. A better name than either is *diode* ("two electrodes"), since two sealed elements—the filament and the plate—serve as electrodes within the bulb.

Two years later, in 1906, the American inventor Lee De Forest (1873–1961) added a third sealed element to the tube and made it a *triode*. The third element consisted of a network of fine wires placed between the filament and the plate. This network is the *grid*.

The grid serves to make the control of the electron flow much more delicate. In the diode, the current either flows or does not flow; the valve is pretty much either wide open or tight shut. The mere mechanical presence of the grid would have little effect on this, for almost all the electrons would slip through the holes. A very small proportion would strike the wires themselves and be stopped.

Suppose, though, that the grid were part of a separate electrical circuit and that a small negative charge were maintained upon it. Each wire of the grid would then repel the electrons, which would be deflected if they came too close. In addition to the mechanical obstruction of the wire itself, each wire would be thickened, so to speak, by a layer of electrical obstruction. The holes through which the electrons could pass without being turned back would become smaller, so that fewer electrons would reach the plate. If the grid were made slightly more negative, the effect would become more pronounced; it would not take much of a negative charge on the grid to cut off the current completely, despite the positive charge on the plate behind the grid. The ordinary valve action could now be allowed to remain wide open while the grid took over control.

The result would be most important if the grid were part of a circuit in which a very weak and varying current was set up. The negative charge on the grid would vary slightly, in perfect step with the variation in current potential, and this variation would open and shut the valve between the filament and the plate. The very small variation in negative potential on the grid would result in a very large variation on the current getting through the grid. The large current would, however, keep exact step with the weak grid potential, and also imitate its variations exactly. The characteristic of a weak current would be imposed on a strong one, and the triode would act as an *amplifier*.

Inventors now had a method of producing effects by altering the motion of tiny, almost massless electrons, rather than by altering the motion of comparatively large and massive levers and gears. The electrons, with so little mass had equivalently little inertia, so that changes could be enforced upon them in tiny fractions of a second. The proverbial "wink of an eye," fast in comparison to the behavior of mechanical devices, became slowness itself in comparison with the rapid action of electronic instruments.

Radio

Diodes, triodes, and various more complicated descendants were put to work in connection with a device even more dramatic than the electric light that gave them birth.

This dated back to the discovery of radio waves by Hertz, who had produced radio waves at one point and had detected them at another. It was easy to imagine that if radio waves could be produced easily enough and detected sensitively enough, the distance between the point of production and the point of detection might be made miles rather than feet. Consequently, if the radio waves were produced in bursts that imitated the Morse code, for instance, a form of communication would be established. The effect of the telegraph (see page II–209) would be duplicated, with radio waves across space replacing electric currents along wires.

The result might be called "wireless telegraphy" or "radio-telegraphy." Actually, the British call it the former, shortening it to "wireless," while Americans call it the latter, shortening it to *radio*.

An Italian electrical engineer, Guglielmo Marconi (1874–1937), having read a description of Hertz's experiment in 1894, set about making communication by way of radio waves a reality. He made use of Hertz's method of producing the radio waves, and of a device called the "coherer" to detect them. The coherer consisted of a container of loosely-packed metal filings. Ordinarily this conducted little current, but it conducted quite a bit when radio waves fell upon it. In this way, radio waves could be converted into an easily detected electrical current.

Gradually Marconi added devices that facilitated both sending and receiving. In 1895, he sent a signal one mile; in 1896, nine miles; in 1897, twelve miles; and in 1898, eighteen miles. He even established a commercial company for the sending of "Marconigrams."

In all this a seeming paradox appeared. Radio waves, like any other form of electromagnetic radiation, ought to travel in straight lines only, and therefore, like light, should be able to penetrate no farther than the horizon. Beyond the horizon, the bulge of the spherical earth should have interfered.

Marconi noted, however, that radio waves seemed to follow the curve of the earth. He had no explanation for this, but he did not hesitate to make use of the fact. On December 12, 1901, Marconi succeeded in sending a radio wave signal from the southwest tip of England, around the bulge of the earth, to Newfoundland. He had sent a signal across the Atlantic Ocean, and this may be taken as a convenient date for the "invention of radio."

Within the year, an explanation for radio communication around the earth's bulge was offered independently by the British-American electrical engineer Arthur Edwin Kennelly (1861–1939) and the English physicist Oliver Heaviside (1850–1925). In the upper atmosphere, both pointed out, there must be regions rich in electrically charged particles. Such particles, both went on to show, would serve to reflect radio waves, which would then cross the Atlantic Ocean, not in a direct curved path, but in a series of straight-line reflections between heaven and earth.

These regions of charged particles were actually detected in 1924 by the English physicist Edward Victor Appleton (1892–1965). In honor of the original theorists, the region is sometimes referred to as the *Kennelly-Heaviside layer*. The charged particles are, of course, ions, and that portion of the upper atmosphere is therefore called the *ionosphere*.

The use of radio waves to make wireless telegraphy possible was only the beginning. Might they not be used to transmit sounds, and not merely pulses? Suppose radio waves could be made to pull a diaphragm in and out and thus set up sound waves in the air?

At first this thought might seem impractical. Radio waves, though far lower in frequency than light waves, are nevertheless far higher in frequency than sound waves. A typical radio wave might have a frequency of 1,000,000 cycles per second (or 1000 kilocycles per second) and it would not be useful to force a diaphragm to vibrate at that frequency. The sound would be far too high-pitched for the human ear to hear. To produce sounds within the range of human hearing, a diaphragm must be made to vibrate between 20 and 20,000 cycles per second. These are the *audio-frequencies*. To use radio waves of such frequencies would be to involve one's self with radiation so low in energy as to be unusable.

The attack was made differently. The radio wave itself was allowed to be uniform and featureless, and with a frequency far above the audio range. It was a *carrier wave*, which served merely to transport the message that was to be impressed on it. Sounds picked up by a microphone could then be used to set up a current that would alter the intensity of the carrier wave in exact step with the fluctuations of the sound waves as in the case of a telephone mouthpiece (see page II–210). This fluctuating current is then made to alter the energy of the carrier wave, the amplitude of which will rise and fall with the rise and fall of the sound waves.

Amplitude modulation

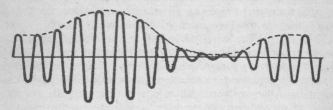

Frequency modulation

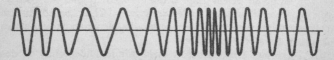

The carrier wave thus regulated is said to be *modulated*. Since the modulation takes the form of alterations in amplitude to match the variability of a sound wave, it is said to be *amplitude modulation*, often abbreviated AM.

When such a modulated radio wave is received, it is first rectified so that only the top half of the wave is allowed through. That half of the wave then acts upon a diaphragm by setting up a fluctuating magnetic force, as in the case of a telephone receiver. The diaphragm cannot react to the rapid fluctuations of the carrier wave itself but only to the much slower variations in its amplitude. In this way, sound waves are reproduced that exactly mimic those that had originally been impressed upon the carrier wave.

In 1906, the Canadian-American physicist Reginald Aubrey Fessenden (1866–1932) first made use of a modulated carrier wave to send out an appropriate message that allowed receivers actually to pick up music. Thus radio meant not only "radiotelegraphy" but also "radiotelephony."

None of this would be truly practical without the use of vacuum tubes for properly manipulating the excessively feeble electric currents set up by radio waves. In fact, so important were these devices to radio that they came to be commonly called *radio tubes*.

Each radio transmitting station makes use of a carrier wave of distinctive frequency. The radio set can be tuned by adjusting a variable condenser (see page II–172) to a point that will allow the set to respond to a particular frequency. In the first two decades of radio, this was not an easy task, and radio enthusiasts had to develop considerable skill at it.

During World War I, however, the American electrical engineer Edwin Howard Armstrong (1890–1954) invented what came to be called a *superheterodyne receiver*. Armstrong had tried to work out a system for detecting airplanes at a distance by picking up the electromagnetic waves sent out by their ignition systems. Those waves were too high in frequency to be received easily; Armstrong therefore arranged to produce a second electromagnetic wave of somewhat different frequency from that which he was trying to detect. The two combined to produce "beats" exactly as sound waves would (see page I–169). The beats were of far lower frequency than either original wave and could easily be detected.

World War I ended before Armstrong could perfect his device, but it was thereafter applied to radio sets in such a way as to make it simple to tune in stations by the turn of a dial. Radio moved into the home in consequence.

In later years, Armstrong tackled another problem involved in radio reception—that of "static." Electromagnetic waves are set up by spark discharges in automobile ignition systems, in the brushes of electric motors, in thermostats, and in all sorts of electrical appliances. (They are also set up in lightning discharges—giant sparks—during thunderstorms.) These waves interfere with the entire range of carrier waves, modulating them in random fashion so that one hears sharp, crackling noises that can be very distracting and cannot be tuned out.

Armstrong devised circuits that modulated not the amplitude of a carrier wave, but its frequency. Such *frequency modulation*, or FM, is not affected by the electromagnetic waves that pulse randomly all about us; consequently, static is largely abolished. In addition, FM allows better reproduction in the extreme portions of the audio-frequency range.

Television and Radar

The cathode-ray tube itself came into direct use in an electronic instrument that was fated to replace radio in the public heart. The beginning here came when physicists learned to take advantage of the low inertia of the electrons in order to move the stream with great rapidity.

Imagine, for instance, a cathode-ray tube with its anode in the form of a hollow cylinder. The electron beam, hurrying in the direction of the anode, would pass through the cylinder to the other end of the tube, which flares out to a flat circular piece of glass coated inside with some fluorescent chemical. Where the electron beam strikes, there would be a brilliant spot of flourescence.

Suppose, though, that on its way to the screen, the electron beam passed between two vertical electrodes. The electron beam will be deflected, naturally, in the direction of the positive electrode. If one electrode carries a strong positive charge to begin with, the electron beam would be strongly deflected in that direction, and the fluorescent spot would appear at the very edge of the screen.

If the positive charge is gradually weakened, the beam's deflection is decreased and the spot moves toward the center of the screen. Eventually, as the positive charge is decreased to zero and as the electrode in question then becomes negative (with the other electrode taking its turn at being positive), the spot passes the center and moves all the way to the other end of the screen. If the maximum positive charge is then placed once more on the first electrode, the beam flashes backward and the spot appears in its original position again.

This can be repeated over and over again, the fluorescent spot drifing across the screen over and over again. This can easily be done quickly enough to cause the spot to become a bright, horizontal line—the eye being unable to see it as a moving spot (It is a similar effect that allows the eye to see the successive stills of a motion picture film as representing moving objects.)

Next imagine the electron beam also passing between a second pair of plates, a pair oriented horizontally This second pair, acting alone, could be used to make the electron beam mark out a vertical line.

If both plates work together, however, the results can be most useful. The first plate may have superimposed upon it the change in voltage required to bring about a steady horizontal line. The second pair of plates may be hooked up to an ordinary alternating current so that the charge on the plates oscillates rapidly and evenly. The action of the two taken together would form a sine wave.

If the current passing through the second pair of plates is made to vary in accordance with a particular set of sound waves, the electron beam would trace out a varying curve that would mimic the properties of the sound wave (translating the longitudinal sound wave into an analogous transverse wave, however— see page I–150) For this reason, when the German physicist Karl Ferdinand Braun (1850–1918) introduced such a device, it came to be called a *cathode-ray oscillograph* ("wave-writer")

The cathode-ray oscillograph can do more. Imagine that the second pair of plates increases its voltage in steps, so that after the electron beam marks out a horizontal line, it moves up a bit and marks out another horizontal line, then moves up and marks out still another, and so on. The entire screen may thus be divided into hundreds of lines, but so fast do voltages shift, and so quickly do the electrons shift with them, that the entire screen can be scanned many times per second. To the eye, then, the entire screen will appear lit up, though a close look will show that the lighting consists of horizontal lines separated by narrow dark spaces that represent the step through which the second pair of plates has lifted the electron beam.

This, in essence, is a *television tube*. In order to impress a picture on it, the electron beam must be strengthened and weakened according to some fixed pattern so that the fluorescent spot is made to grow brighter and dimmer, producing the light-dark pattern we would recognize as an image.

The first to produce a practical method for doing this was the Russian-American physicist Vladimir Kosma Zworykin (1889–) In 1938, he invented the *iconoscope* (from Greek words meaning "picture-viewer") It was a camera of sorts, one in which the rear surface was coated not with photographic film, but with a large number of tiny droplets of an alloy of cesium and silver. Cesium readily gives off electrons when light falls upon it, the intensity of electron emission being proportional to the intensity of the light that falls upon it. When the light-dark pattern of the scene in front of the camera is focussed on the cesium-silver rear surface, an analogous pattern of many-electrons/few-electrons is produced.

This electron pattern can be made to influence the electron beam emitted in the television tube which causes the fluorescent spot on the television screen to brighten and dim in exact analogy to the light-dark pattern being viewed by the iconoscope. The entire picture is reproduced on the screen; and since this is done over and over many times per second, each time in a slightly different pattern (as the scene being viewed changes), the eye seems to make out motion.

The cathode-ray oscillograph is also used in connection with a device that makes use of electromagnetic waves to judge distance, much as sound waves are used in echo location (see page I–180).

Electromagnetic waves move at the precisely known, very high velocity of 300,000 kilometers per second. Imagine a short pulse of electromagnetic waves moving outward, striking some obstacle, and being reflected backward and received at the point from which it had issued forth an instant before. What is needed is a wave form of low enough frequency to penetrate fog, mist and cloud, but of high enough frequency to be reflected efficiently. The ideal range was found to be in the microwave region, with wavelengths of from 0.5 to 100 centimeters.

From the time lapse between the emission of the pulse and the return of the echo, the distance of the reflecting object can be estimated. And, of course, the direction of the reflecting object would be that in which reflection was sharpest.

A number of physicists worked on devices making use of

this principle, but the Scottish physicist Robert Alexander Watson-Watt (1892–   ) was the first to make it thoroughly practical. By 1935, he had made it possible to follow an airplane by the microwave reflections it sent back. The system was called "radio detection and ranging" (to "get a range" on an object is to determine its distance), and this was abbreviated to "ra. d. a. r." or *radar*.

The microwave pulse sent out in radar can be made to deflect the electron beam of a cathode-ray oscillograph upward, producing a sharp spike in what would otherwise be a horizontal line. The returning echo (much feebler than the original pulse, since only a portion of the pulse strikes the object it is aimed at, and some of the pulse that does strike is scattered in other directions) produces a smaller spike. The electron beam moves sideways with such rapidity that even though the echo arrives only a fraction of a millisecond after the pulse has been sent out, there is still ample space on the fluorescent line between pulse and echo—a space that can be measured and made to yield distance.

Another way in which an electron beam can be made to do this work is to have it start at the center of the screen and move out to the edge along any radius. The radius it chooses is governed by the direction in which the large radar antenna (designed to receive and magnify feeble echoes) is pointing. As the antenna makes a complete circle, the electron beam very rapidly sweeps out a series of radii all around the screen.

Returning echoes make themselves evident not by sharp deviations in the beam itself, but by a brightening of the beam intensity; consequently, an obstructing object, returning echoes, shows up as a bright spot on the screen. If the screen is coated with a

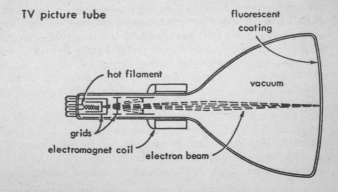

TV picture tube

fluorescent coating

hot filament

vacuum

grids

electromagnet coil

electron beam

substance whose fluorescence lingers a few seconds, the shape of the object may be roughly scanned out as the beam completes its sweep. From an airplane, the radar screen may even present a rough map of the ground below, since land, water, green leaves and concrete all reflect microwaves at differing intensity

It is not only man-produced microwaves that can now be detected by the help of electronic instruments. The various heavenly bodies and the phenomena with which they are associated produce among themselves the entire range of the electromagnetic spectrum. Little of that spectrum, however, can penetrate the earth's atmosphere. Among that little, fortunately, is the visible light region in which our sun's radiation happens to be particularly rich.

Another region, however, to which the atmosphere is transparent is that in which the microwaves are found.

In 1931, the American radio engineer Karl Jansky (1905–1950) was engaged in the problem of tracking down causes of static. Having eliminated static caused by known disturbances, he found a new kind of weak static from a source which, at first, he could not identify. It came from overhead and moved steadily from hour to hour. At first it seemed to Jansky that the source moved with the sun. However, it gained slightly on the sun to the extent of four minutes a day. Since this is just the amount by which the vault of the stars gains on the sun, the source must lie somewhere among the stars outside the solar system.

By 1932, Jansky had decided the source was strongest in the direction of the constellation of Sagittarius—in which direction, astronomers had decided, lay the center of the Galaxy

The center of the Galaxy, hidden from optical view by dust clouds that efficiently absorb all the light, is nevertheless apparent through its microwave emissions which penetrate the dust clouds. *Radio telescopes* were built to receive and focus the very weak signals (especially after World War II when advances in radar technology could be put to this use), and the new science of *radio astronomy* came into its own.

4

Electrons Within Atoms

The Photoelectric Effect

For a brief while after the discovery of the electron, it might have been tempting to feel that the universe contained at least two sets of ultimate particles without necessary connection with each other. One set consisted of the atoms of matter, these being comparatively massive objects existing in dozens of varieties. The other set consisted of the electrons associated with an electric current which, to all appearances, came in but a single variety.

Yet there was reason for doubting the independence of these two sets of particles. When an electric current was first produced by Volta a century before the discovery of the electron, it was done by combining certain metals and solutions. Since that time any number of chemical cells—devices whereby an electric current originates as a result of some chemical reaction—were devised The ordinary "flashlight battery" and the storage battery present in every automobile are the best-known examples of these.

If a group of chemicals, each electrically neutral when taken by itself, could give rise to electric current made up of myriads of electrons, then certainly the worlds of atoms and of electrons must have some connection. Furthermore, one had to believe that either the electrons of the current were formed in the process of the atomic joinings and atomic separations that make up a chemical reaction, or that the electrons were present in the chemicals at all times, and were merely released in the course of the reaction.

Both views had their difficulties. If electrons were formed, that meant that mass was created, and that seemed impossible in the light of the law of conservation of mass (see page II–107), a generalization which, during the 1890's, was completely accepted by scientists. On the other hand, if electrons were present in chemicals at all times, why was there ordinarily no evidence either of their existence or, particularly, of the electric charge associated with them?

The dilemma was made the more acute through a phenomenon that was already known to physicists at the time of the discovery of the electron.

When Hertz was experimenting with radio waves during the 1880's, he found that he could elicit a spark from his radio-wave detector more easily if light fell upon the metal points giving out the spark. Light drew electricity from metal, so to speak, and this came to be called the *photoelectric effect*.

In 1888, the German physicist Wilhelm Hallwachs (1859–1922) discovered that light affected the two varieties of electric charge differently. A negatively-charged zinc plate lost its charge if it was exposed to ultraviolet light, the charge being drawn out by the light. On the other hand, a positively-charged zinc plate was not affected by the ultraviolet light.

Once the electron was discovered, a reasonable explanation of the phenomenon at once offered itself. Those were electrons that were ejected from metal through the impact of light. It was these electrons that formed the spark. It was from a negatively-charged zinc plate, containing an excess of electrons, that those particles were easily ejected. Such particles were not ejected from a positively-charged plate, which clearly did not contain an excess of electrons.

In 1899, Thomson tested this notion by measuring the e/m ratio for the particles being ejected from metals under the influence of light; it turned out to be virtually identical with that of cathode-ray particles. They were accepted as electrons from that time.

Again the same problem arose. When light forced electrons out of an electrically-neutral metallic surface, were those electrons formed as they were emitted or did they exist within the metal at all times? By 1905, Einstein had shown that the law of conservation of mass was incomplete in the form that had generally been accepted during the nineteenth century. He showed that energy and mass could be interconverted and that one ought to speak of the law of conservation of mass-energy. Nevertheless, the book-keeping involved in the interconversion of mass and energy was rigorous, and there was insufficient energy in ordinary light—and even in ultraviolet light—to serve the purpose of manufacturing electrons.

Electrons, then, must exist in the metal at all times, and one could ask another question. Did the electrons exist in the interstices between the atoms, or did they actually occur within the atoms themselves? It was hard to accept the latter view, for that would mean that the atom was not the featureless, ultimately indivisible object that Democritus and Dalton had proposed and that the scientific world had finally accepted.

Yet there were phenomena that seemed to make this anti-Democritean view necessary, perhaps. Philipp Lenard had observed that the energy with which electrons were ejected depended on the frequency of the light, and that light of less than a certain frequency (the *threshold value*) did not eject electrons. The quantum theory (see page II–130), which was beginning to come into acceptance in the first decade of the twentieth century, made it clear that light consisted of photons that increased in energy content as frequency increased.

The threshold value represented quanta of just sufficient energy to break the bonds holding the electrons to matter. The strength of those bonds varies from substance to substance, since electrons are forced out of some metals only by energetic ultraviolet light, whereas they are forced out of other metals by light as un-energetic as the visible red. If electrons are tied to matter, it must be to the atoms they are bound, and with differing bond strengths, so to speak, depending on the nature of the particular atom. It seems only sensible to consider something always present near the atom, always bound to the atom with a characteristic force, to be part of the atom.

Furthermore, once the view is accepted, there are advantages to it. There are many varieties of atom and only one type of electron (since the particles emitted from all metals by the photoelectric effect are of identical properties). Perhaps the troublesome variety of the atoms could be explained in terms of the number of electrons each contained, of their arrangement, of the strength with which they were held, and so on. Perhaps the order enforced empirically upon the elements by the periodic table could now be made more systematic. If so, the indivisible atom of Democritus was well lost.

Indeed, there were some facets of the photoelectric effect that fit in well with the periodic table. For instance, the elements that most readily give up electrons in response to light are the alkali metals. These give up electrons with increasing ease as atomic weight goes up—that is, as one moves down the column in the periodic table. Thus cesium, the naturally-occurring alkali metal with the highest atomic weight,* releases its electrons most easily of all—hence Zworykin's use of the metal in his iconoscope.

Here is an indication of how Mendeleev's periodic table established a kind of order with respect to a property completely undreamed of in Mendeleev's time. This is an example of how a truly useful scientific generalization can be superior to the state of knowledge that brought it forth, and how a great scientist must almost necessarily produce more than he realizes.

The photoelectric effect can be put to good use. A vacuum tube can be devised that does not require a heated filament for the production of electrons—merely a filament (if one chooses the right metal) that can be exposed to light. When light falls upon a cathode capable of showing a photoelectric effect in response to such light, electrons are ejected and a current flows. The current can be used to activate an electromagnet that can open doors or perform other tasks. This is a *photoelectric cell*.

A common version of such a cell places it in one post with a source of light from another post shining constantly into the cell, keeping a current constantly flowing and a door constantly closed against a pull that would otherwise open it. A person walking between the posts intercepts the light beam, the current in the cell ceases, and the door flies open.

The Nuclear Atom

The apparent existence of electrons within the atom raised some important questions.

The atoms were electrically neutral; if negatively-charged electrons existed about or within the atom, there had to be a positive charge somewhere to neutralize the negative charge of the electrons. If so, where was it? Why didn't light ever bring about the ejection of very light positively-charged particles? Why were there only cathode rays, never analogous anode rays?

* Francium, an alkali metal of still higher atomic weight, does not occur in nature in any significant quantity (see pag 130).

Thomson offered an answer to these questions. In 1898, he suggested that the atom was a solid, positively-charged sphere into which just enough electrons were embedded (like raisins in pound cake, so to speak) to bring about an overall electrical neutrality.

This was an attractive suggestion, for it seemed to explain a great deal. Light quanta would jar loose one or more of these electrons, but could scarcely budge the large atom-sphere of positive charge. Again, the heat in a vacuum tube filament would indeed "boil off" electrons, for as atoms vibrated more strongly with rising temperature (in accordance with kinetic theory, see page I-205), the electrons would be jarred loose while the atom itself would be essentially unaffected. This would explain why only negative particles appeared and never positive ones.

Then, too, Thomson's theory explained ions neatly An atom that lost one or more electrons would retain a net positive charge —the size of the charge depending on the number of electrons lost. A hydrogen ion (H^+) or a sodium ion (Na^+) would be a hydrogen atom or a sodium atom that had lost a single electron. A calcium ion (Ca^{++}) would be a calcium atom minus two electrons, and an aluminum ion (Al^{+++}) would be an aluminum atom minus three electrons.

On the other hand, what if more than the normal quantity of electrons could be jabbed into the positively-charged atom substance? The chloride ion (Cl^-) would be a chlorine atom bearing an extra electron, while a sulfate ion (SO_4^{--}) and a phosphate ion (PO_4^{---}) would represent groups of atoms possessing among themselves two and three extra electrons, respectively.

In this view, the negatively-charged electron is the only subatomic particle, but by means of it ions of both types of electric charge can be explained.

Thomson's theory, although so attractive, nevertheless, had a fatal shortcoming. Lenard had noted that cathode rays could pass through small thicknesses of matter To be sure, the electrons making up the cathode rays were very small and might be pictured as worming their way between the atoms. If so, they would most likely emerge badly scattered. Instead, cathode rays passed through small thicknesses of matter still traveling in an essentially parallel beam, as though they had passed through atoms without very much interference.

In 1903, therefore, Lenard suggested that the atom was not a solid mass but was rather mostly empty space. The atom, in his view, consisted of tiny electrons and equivalent particles of positive charge, existing in pairs so that the atom as a whole was electrically neutral.

But, in that case, why were there only cathode rays and never anode rays?

The reconciliation of the Thomson and Lenard views fell to the lot of the New Zealand-born physicist Ernest Rutherford (1871–1937). Beginning in 1906, he conducted crucial experiments in which he bombarded thin gold leaf with alpha particles.* Behind the gold leaf was a photographic plate.

The stream of alpha particles passed right through the gold leaf as though it were not there and fogged the photographic plate behind it. The gold leaf was only 1/50,000 of a centimeter thick, but this still meant a thickness of 20,000 atoms. The fact that alpha particles could pass through 20,000 gold atoms as though they weren't there was strongly in favor of Lenard's notion of an empty atom, (an atom, that is, made up of nothing more than a scattering of light particles).

But the truly interesting point was that not all the alpha particles passed through unaffected. The spot of fogging on the plate would, in the absence of the gold leaf, have been sharp; but with the gold leaf in place, the boundary of the fogged spot was rather diffuse, fading out gradually. It was as though some of the alpha particles were, after all, slightly deflected from their path. In fact, Rutherford was able to show that some were deflected more than slightly! About one alpha particle out of every 8000 was deflected through a right angle or even more.

This was amazing. If so many alpha particles went through thousands of atoms untouched or nearly untouched, why should a very few be twisted in their path so badly? The alpha particle is not a light particle such as the electron. It is 7350 times as massive as an electron; four times as massive as a hydrogen atom. If the alpha particle encountered electrons within an atom, it would brush them aside as a man might brush aside a sparrow. For an alpha particle to be set back on its heels, it must at the very least meet something nearly as massive as itself—something, in short, of atom-sized mass. And yet this atom-sized mass was only rarely encountered on the journey of the alpha particle through matter, so it must take up a very small volume.

It was as though one were faced with contacting fluffy balls of foam with a lead pellet at the center of each. If lead pellets were

---

* Alpha particles are rapid-moving, massive particles obtained from radioactive substances (see page 111) capable of penetrating matter with much greater effectiveness than electrons can.

tossed at such a barrier, most would pass through the foam as though nothing were there, but occasionally a tossed pellet would strike one of the buried pellets and bounce off. From the frequency with which such bouncing took place, you could calculate the comparative size of the foam ball and the central pellet.

To be sure, the alpha particles were not actually bouncing off the massive object within the atom. Instead, from the nature of the scattering, Rutherford could show there was an electrical interaction. The alpha particles are themselves positively charged (each carrying a charge of +2), and the massive object within the atom is also charged (positively, as it turns out), so the alpha particle is repelled by electric forces even if it scores a near miss.

By 1911, Rutherford was ready to describe his picture of the atom. In his view, Thomson's massive positively-charged atom was still there as far as mass was concerned, but it was drastically shrunken in volume. It had shrunk down to an extremely small object in the very center of the atom. This massive central object was the *atomic nucleus,* and what Rutherford was proposing was the *nuclear atom,* a concept that has remained valid ever since and that is more firmly accepted now than ever.

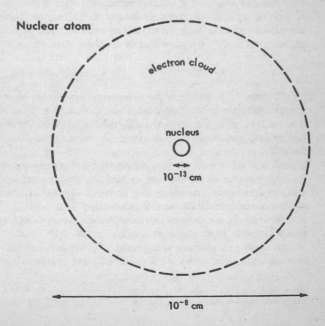

Nuclear atom

electron cloud

nucleus

10^{-13} cm

10^{-8} cm

The atomic nucleus, as could be seen from the pattern of deflections of alpha particles, was tiny indeed, not more than 10^{-13} to 10^{-12} centimeters in diameter, or only 1/100,000 to 1/10,000 the diameter of the atom as a whole. The volumes of nucleus and atom are in proportion to the cube of the diameter, so the volume of the nucleus is rather less than one trillionth (1/1,000,000,000,-000) of the atom as a whole.

Yet virtually all the mass of the atom is concentrated in that tiny nucleus. Even the lightest nucleus, that of the hydrogen atom, is 1836 times the mass of an electron, while the nuclei of the really massive atoms are nearly half a million times as massive. Such a nucleus would be much less mobile than electrons would be, and it is not surprising that light ejects negatively-charged electrons and not positively-charged nuclei from metals—that heated filaments emit electrons and not nuclei.

Outside the nucleus, the comparatively vast remainder of the atom is made up of nothing but the ultra-light electrons. These electrons offer little obstacle to speeding cathode ray particles, and virtually no obstacle at all to alpha particles; consequently, Rutherford's nuclear atom is as thoroughly empty as Lenard's model was.

And, of course, the nuclear atom can explain ions in terms of loss or gain of electrons as easily as Thomson's raisin-cake atom could explain them. In short, the nuclear atom proved completely satisfactory; only the details required elaboration.

Characteristic X Rays

Thanks to Rutherford, physicists now saw the atom as a tiny but massive, positively-charged nucleus surrounded by electrons. The nucleus, if it contained virtually all the mass of the atom, as Rutherford maintained, must vary in mass with the atomic weight.

It seemed reasonable to suppose that the greater the mass of the nucleus, the larger the size of the positive charge it carried and the greater the number of negatively-charged electrons necessarily present outside the nucleus to balance that positive charge. If this were so, it would mean that physicists were beginning to probe close to what might prove the crucial difference between the atoms of one element and another. It was not just the difference in mass, which was all that Dalton, and nineteenth century chemists generally, could put their finger on. A possible new difference was emerging, an electrical difference that made itself manifest in two ways: first, in the size of the positive charge on the nucleus and second, in the number of electrons outside the nucleus.

These two aspects of the electrical differences among atoms are closely related, but the nuclear charge is more fundamental than the electron number. Electrons can be removed from atoms by heat or by light, leaving positive ions behind. Additional electrons can be forced onto atoms in chemical reactions, forming negative ions. While these ions have properties that differ radically from those of the neutral atom, they are not completely divorced from the neutral atom; they do not constitute a new element. In other words, the sodium ion is very different from the sodium atom, but one can be changed into the other by recognized nineteenth century chemical or physical procedures. Neither can be changed into either a potassium atom or a potassium ion, at least not by those procedures. Therefore, changes in the electron number in an atom are not necessarily crucial, and it is not by means of the number of electrons within an atom that elements are best distinguished.

On the other hand, the nuclear charge could not be altered by any method known to the chemists and physicists of 1900; no alteration of the number of electrons, one way or the other, would alter that nuclear charge. It was the size of the nuclear charge, then, that best characterized the different varieties of atoms and, therefore, the different elements.

But if all this is so, how can one go about finding the exact size of the nuclear charge of the atoms of a particular element? The answer to that question was arrived at through X rays.

When Röntgen first discovered X rays, he had produced them as a result of the impact of cathode ray particles on the glass at the end of the cathode-ray tube. Speeding electrons can penetrate small thicknesses of matter, but they are slowed down; if the obstructing matter is thick enough, they are stopped completely and absorbed. The deceleration of electrically charged particles will, according to Maxwell's theory of electromagnetism, result in the production of electromagnetic radiation, and this does indeed appear in the form of X rays.

It is to be expected that material made up of massive atoms will more effectively and rapidly decelerate speeding electrons and will produce more intense beams of X rays. For that reason, physicists took to placing metal plates directly opposite the cathode inside cathode-ray tubes. This metal plate, sometimes called the *anticathode* ("opposite the cathode") is subjected to the collision of electrons, and from its surface, powerful beams of

X rays are emitted. Such a cathode-ray tube is usually called an *X-ray tube*.

The X rays produced from the anticathode varied in properties according to the nature of the material making up the anticathode. The first to show this was the English physicist Charles Glover Barkla (1877–1944). In 1911, Barkla showed that among the X rays produced at a given anticathode, certain groups predominated. He could only judge the difference among the X-ray groups produced by their ability to penetrate thicknesses of matter. One group would penetrate a relatively large thickness, another group a lesser thickness, and so on. The greater the thickness penetrated, the "harder" the X rays. It became customary to call the hardest X rays produced at a given anticathode, the *K-series*, the next the *L-series*, then the *M-series*, and so on. These are the *characteristic X rays* for a given element.

The hardness of these sets of characteristic X rays varies with the nature of the metal making up the anticathode. In general, the higher the atomic weight of the metal, the harder the X rays produced. It seemed reasonable to suppose that if the hardness could be measured accurately, interesting information concerning the atomic nuclei could be obtained.

Unfortunately, measuring the hardness of X rays by their penetrability is rather imprecise. Something better was needed. It was strongly suspected, for instance, that X rays were electromagnetic radiation (though when Barkla did his work this had not yet been conclusively demonstrated). If so the shorter the wavelength of a particular beam of X rays, the more energetic it would be and the more penetrating. Measurement of the wavelength of the X rays (or of their frequency) would thus offer a possibly precise method for estimating their hardness.

However, how could their wavelength be measured? In principle, the best method would be to use a diffraction grating (see page II–65). A diffraction grating, made of a series of parallel scratches on an otherwise clearly transmitting surface, can only work under certain circumstances. The distance between scratches must approximate the size of the wavelength being measured. The wavelength of X rays was far shorter than that of ultraviolet radiation, so short, in fact, that it was impractical to expect scratches to be produced with sufficiently close spacing.

A way out of this dilemma occurred in 1912 to the German physicist Max von Laue (1879–1960). Crystals, he realized, were natural diffraction gratings far more finely spaced than any man could make. In crystals, atoms existed in orderly rows and files.

The nuclei of the atoms, which would deflect X rays just as scratches would deflect ordinary light, are about 10^{-8} centimeters apart (this being roughly the diameter of a typical atom), and this might very well be about the size of an X ray wavelength.

Laue used a crystal of zinc sulfide, allowing a beam of X rays to fall upon it and, passing through, to strike a photographic plate. The X rays were indeed diffracted, producing a pattern of dots, instead of a single, centrally located dot. This was the definite proof, at last, that X rays were wave-like in nature.

This approach was carried further that same year by a pair of physicists, the Englishman William Henry Bragg (1862–1942) and his son, the Australian–born William Lawrence Bragg (1890–   ). They analyzed the manner in which X rays would be reflected by the planes of atoms within a crystal, and showed that this reflection would be most intense at certain angles, the values of which depended upon the distance between the planes of atoms within a crystal and upon the wavelength of the X ray. If the distance between the planes of atoms was known, the wavelength could then be calculated.

It was found that by this method the wavelength could be calculated with a satisfactory degree of precision. X rays, produced by the deceleration of speeding electrons, have been found across the entire range of from 1 millimicron (the arbitrary lower limit of ultraviolet radiation wavelengths, see page 34) down to somewhat less than 0.01 millimicrons, a range of about seven octaves.

Atomic Numbers

With the Bragg technique at hand, it was now possible to turn to Barkla's characteristic X rays and study them carefully and precisely. This was done in 1913 by the English physicist Henry Gwyn-Jeffreys Moseley (1887–1915).

Moseley worked with the K-series of characteristic X rays for about a dozen consecutive elements in the periodic table, from calcium to zinc, and found that the wavelength of the X rays went down (and the frequency therefore went up) as the atomic weight increased. By taking the square root of the frequency, he found that there was a constant increase as one went from one element to the next.

Moseley decided that there was something about the atom which increased by regular steps as one went up the periodic table. It was possible to demonstrate that this "something" was most likely the positive charge on the nucleus. The most straight-forward conclusion Moseley could reach was that the simplest atom had a charge of +1 on its nucleus; the next, a charge of +2; the next, a charge of +3, and so on. Moseley called the size of this charge the *atomic number*.

This has turned out to be correct. Hydrogen is now considered to have an atomic number of 1; helium, one of 2, lithium, one of 3, and so on. In the periodic table presented on page 16, the elements are given atomic numbers of from 1 to 103, and the atomic number of every known element has been determined.

The atomic number is far more fundamental to the periodic table than the atomic weight is. Mendeleev had been forced to put some elements out of atomic weight order so that they would fit into their proper families. For instance, cobalt fits the table better if it is placed ahead of nickel. Yet cobalt, with an atomic weight of 58.93, should fall behind nickel, which has an atomic weight of only 58.71.

Moseley, however, found that cobalt, despite its heavier atomic weight, produced X rays that were lower in frequency than those of nickel. Cobalt therefore has the lower atomic number, 27, and the atomic number of nickel is 28. Mendeleev's chemical intuition, working without the guide of X-ray data, had led him aright.

To summarize, there are three pairs of elements in the periodic table (argon-potassium, cobalt-nickel, and tellurium-iodine) which are out of order if increasing atomic weight is taken as the criterion. If increasing atomic number is taken as the criterion instead, not one element in the table is out of order.

The atomic number concept also brought new power to the periodic table in another way. Not only could chemists predict missing elements (as Mendeleev had), but they could now also predict the nonexistence of elements.

As long as atomic weight had been the only guide, one could never be certain that whole new families of undiscovered elements might not exist. In the 1890's, for instance, the family of inert gases—helium, neon, argon, krypton, and xenon—was discovered and fitted into a new column in the periodic table, a column no one had previously suspected of existing. Again, the lanthanides were discovered one by one over the space of a century, and until Moseley's time no chemist could be certain how many remained to be found—thousands, for all one could then tell.

With atomic numbers, such uncertainties were smashed. As long as one could assume the nonexistence of fractional electric charges on the nucleus, one could be sure there were no unknown elements between hydrogen (atomic number 1) and helium (atomic number 2), or between phosphorus (atomic number 15) and sulfur (atomic number 16).

In fact, for the first time chemists could tell how many elements remained to be discovered. The first element in the periodic table was hydrogen (atomic number 1), and there could be no element preceding it. The element with the most massive known atom (in Moseley's time) was uranium (atomic number 92). Between these two limits, all the atomic numbers but seven were filled, and only seven unknown elements therefore remained to be discovered. The seven gaps were those with atomic numbers 43, 61, 72, 75, 85, 87, and 91.

X-ray analysis could also be used to check the identity of possibly newly-discovered elements. For instance, the French chemist Georges Urbain (1872–1938) had in 1911 isolated what he thought was a new element, and he had named it "celtium." When Moseley's work was published, Urbain decided his new element must fit into the gap in the periodic table at number 72 and brought a sample to Moseley for testing. Moseley analyzed the characteristic X rays and found the "new element" to be a mixture of ytterbium and lutetium (elements number 70 and 71) both of which were already known. Painstaking chemical work confirmed this and Urbain, very impressed, labored mightily to popularize the concept of the atomic number.

Within a dozen years, three of the gaps were filled. Protactinium (atomic number 91) was discovered in 1917; hafnium (atomic number 72), in 1923; and rhenium (atomic number 75), in 1925. After that, over a decade passed before the last four gaps (43, 61, 85, and 87) were filled. These last elements will be taken up in due course (see page 175).

Once the nuclear charge of an element was known, something was also known about the number of electrons in the atoms of that element. An element might lose an electron or two, or gain an electron or two, and become an electrically-charged ion, but in the *neutral atom*, the number of electrons had to be precisely enough to neutralize the nuclear charge. If, in the oxygen atom, the nucleus has a charge of $+8$, there must be eight electrons (each with a charge of -1) to balance that. We may say, then, that the number of electrons in a neutral atom is equal to the atomic number of the element. The neutral hydrogen atom possesses 1 electron, the neutral sodium atom possesses 11 electrons and the neutral uranium atom possesses 92 electrons.

Electron Shells

The next general question was this: How are the electrons in an atom arranged? Thomson, in his raisin-cake model of the atom, had suggested that the electrons embedded in the positively-charged substance of the atom were arranged in circles. If there were a large number of electrons, there might well be a number of circles.

After Thomson's model had been abandoned and replaced by Rutherford's nuclear atom, it remained possible that the electrons possessed some regular arrangement outside the nucleus. This notion seemed to be backed by the several series of characteristic X rays produced by various elements. Perhaps each series was produced by a separate group of electrons enclosing the central nucleus. The group nearest the nucleus would be most firmly held and would produce the hardest X rays, the K-series. The next group would produce the L-series, and so on. If the electrons were pictured as arranged spherically about the nucleus (like the shells making up an onion), one could speak of the *K-shell*, the *L-shell*, the *M-shell*, and so on, as one worked outward from the nucleus.

Then, consider the inert gases—helium, neon, argon, kryp-

Electron shells

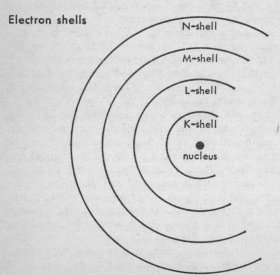

N-shell

M-shell

L-shell

K-shell

nucleus

ton, xenon and radon. Of all the elements, they were the least apt to engage in chemical reactions. (Until 1962, it was taken for granted they could not engage in any chemical reactions at all. It was then discovered that krypton, xenon and radon could engage in a very few.) Why is this so?

One reason is that chemical reactions must involve the interactions of the electrons within atoms. For instance, when sodium metal reacts with gaseous chlorine, sodium chloride is formed and this consists of sodium ions and chloride ions. In the reaction of sodium and chlorine, then, the sodium atom loses an electron to become Na^+, and the chlorine atom gains an electron to become Cl^-.

Perhaps, if the inert gases do not easily engage in chemical reactions, this is because their atoms already possess a particularly stable arrangement of electrons and have only the most minor tendency to upset that arrangement by indulging in the loss or gain of electrons.

It seemed logical to suppose that this stable arrangement is represented by the complete filling of a particular shell of electrons.

For instance, helium has an atomic number of 2, and is inert. One can assume that since the neutral helium atom possesses two electrons, it requires but two electrons to fill the innermost shell, or K-shell. The next inert gas is neon, which has an atomic weight of 10 and which, in its neutral state, possesses ten electrons in its atoms. With two electrons filling the K-shell, the remaining eight must suffice to fill the L-shell. The next inert gas is argon, which has an atomic number of 18, and has eighteen electrons per atom. With two electrons in the K-shell and eight in the L-shell, the remaining eight must fill the M-shell. Based on this reasoning, Table IV contains the distribution of the electrons among the shells of the first twenty elements. (Only the first twenty are included because the distribution becomes more complicated—see page 83—for the higher elements.)

Soon after Moseley's work, an attempt was made to rationalize the chemical reactions on the basis of electron distributions inside the atom. A relatively successful attempt was made, independently, by the American chemists Gilbert Newton Lewis (1875–1946) and Irving Langmuir (1881–1957). The essence of their views was that in any chemical reaction an element gained or lost electrons in such a way as to gain an "inert-gas configuration," that being the most stable arrangement.

Thus, sodium, with its electrons divided 2/8/1, had a strong tendency to give up one electron and become sodium ion (Na^+)

with its electrons divided 2/8. The sodium ion has the neon configuration of electrons but, of course, does not actually become neon, for the nuclear charge of the sodium ion (the characteristic property of a particular element) remains $+11$, while that of neon is $+10$. The same argument will hold for chlorine. The chlorine atom, with an electron arrangement of 2/8/7, has a strong tendency to gain an electron and form the chloride ion (Cl^-), which has a 2/8/8 arrangement, like that of argon.

The ease with which sodium and chlorine interact can be viewed as the consequence of the manner in which their electron shifting tendencies complement each other. The electron that a sodium atom will so easily give up will be accepted just as easily by a chlorine atom. The oppositely charged ions that result cling together to make up sodium chloride.

In the same way, calcium (2/8/8/2) will easily give up two electrons to form calcium ion (Ca^{++}) with a 2/8/8 configuration, that of argon; while oxygen (2/6) will readily accept two

TABLE IV—*Electron Arrangements*

Element	Atomic Number	Electrons in			
		K-shell	L-shell	M-shell	N-shell
Hydrogen	1	1	—	—	—
Helium	2	2	—	—	—
Lithium	3	2	1	—	—
Beryllium	4	2	2	—	—
Boron	5	2	3	—	—
Carbon	6	2	4	—	—
Nitrogen	7	2	5	—	—
Oxygen	8	2	6	—	—
Fluorine	9	2	7	—	—
Neon	10	2	8	—	—
Sodium	11	2	8	1	—
Magnesium	12	2	8	2	—
Aluminum	13	2	8	3	—
Silicon	14	2	8	4	—
Phosphorus	15	2	8	5	—
Sulfur	16	2	8	6	—
Chlorine	17	2	8	7	—
Argon	18	2	8	8	—
Potassium	19	2	8	8	1
Calcium	20	2	8	8	2

electrons to form oxide ion (O^{--}) with a 2/8 configuration, that of neon. Thus, calcium oxide (CaO) is formed.

Or calcium may give up one electron to a chlorine atom and a second electron to another chlorine atom to form calcium chloride ($CaCl_2$). In this way, calcium combines with two chlorine atoms, so that a gram-atomic weight of chlorine combines with only half a gram-atomic weight of calcium. The existence of equivalent weights (with the equivalent weight of calcium being only half its gram-atomic weight) can thus be explained electronically.

Any such theory had to explain why it is that two chlorine atoms cling together tightly to form a chlorine molecule. Each chlorine atom has a strong tendency to accept one electron, but virtually no tendency to give one up. The Lewis–Langmuir suggestion was that each of two chlorine atoms might contribute an electron to a "shared pool" of two. These two electrons would be within the outermost electron shell of both atoms (provided they were so close to each other as to be in virtual contact), and each atom would then have the 2/8/8 configuration of argon.

Anything that would pull the chlorine atoms apart would disrupt this stable electron arrangement by making the existence of the shared pool impossible. For this reason, the two-atom chlorine molecule is very stable, and considerable energy is required to decompose it to individual chlorine atoms.

Similar arguments will explain why fluorine, hydrogen, oxygen and nitrogen all form stable two-atom molecules.

The electron configuration of the carbon atom is 2/4. It may contribute one electron to form a shared pool of two electrons with a hydrogen atom, contribute a second electron to a second hydrogen atom, and so on. In the end, there will be four shared pools, of two electrons each, with each of four hydrogen atoms. The carbon atom shares in eight electrons altogether, four of its own and one each from the four hydrogen atoms, to achieve the 2/8 neon configuration. Each hydrogen atom possesses a share in two electrons to achieve the helium configuration. Thus, the molecule of methane (CH_4) is stable.

Indeed, the Lewis-Langmuir picture of electrons being transferred and shared has turned out to be a very useful way of picturing how the molecules of a great many of the simpler chemical compounds are held together.

Furthermore, the Lewis-Langmuir theory made it plain why the periodic table was periodic (something which Mendeleev, of course, had been unable to explain). To begin with, the inert gases all have their electrons arranged in a way that yields them maximum stability. They are all chemically inert, therefore, and form a natural chemical family of very similar elements.

The alkali metals are all located in positions one atomic number higher than the inert gases. Thus lithium (one past helium) has the electron configuration 2/1; sodium (one past neon) is 2/8/1; potassium (one past argon) is 2/8/8/1, and so on. Every alkali metal has but one electron in its outermost shell and has a strong tendency to lose that one. For that reason, all are very active elements, with similar properties, forming a natural family.

The alkaline earth elements form a similar family in which each has two electrons in the outermost shell of the atom. Beryllium is 2/2, magnesium is 2/8/2, calcium is 2/8/8/2, and so on.

Again, the halogens all are to be found one atomic number before the inert gas configuration. Fluorine is 2/7, chlorine is 2/8/7, and so on. All have a strong tendency to accept one electron and they also form a natural family of elements of similar chemical properties.

And thus, by way of electrons and electron shells, the periodic table was rationalized a half-century after its inception.

CHAPTER 5

Electrons and Quanta

Spectral Series

Useful as the Lewis-Langmuir view of the atom is in explaining the structure of many of the simpler chemical compounds, it does not explain everything. It does not, for instance, describe in satisfactory manner the structure of the boron hydrides (compounds of boron and hydrogen) or explain the peculiar properties of the well-known compound, benzene (C_6H_6). Furthermore, it does not adequately explain the behavior of many of the elements with atomic weights beyond that of calcium. The Lewis-Langmuir view does not, for instance, explain why the lanthanides, with atomic numbers of from 57 to 71 inclusive, should be so similar in properties.

One obvious shortcoming in the Lewis-Langmuir view is that it considers the electrons to be stationary particles distributed about the atom in certain fixed positions. Indeed, the eight electrons of the L-shell and the M-shell were usually depicted as being located at the eight corners of a cube, so that simple molecules could be presented in diagrams as being made up of interlocking cubes.

This is a convenient picture from the chemical point of view, but it is unacceptable to physicists and must be replaced by something else if the Lewis-Langmuir view is to be made more useful. After all, if the negatively-charged electron is stationary with respect to the positively-charged nucleus, then electromagnetic theory requires that it fall into the nucleus (just as the earth would fall into the sun if it were stationary with respect to the sun).

Consequently, physicists tended to assume that the electrons were circling the nucleus at great velocity in order not to fall into it. In 1904, a Japanese physicist, Hantaro Nagaoka, specifically suggested that electrons circled in orbits within the atom, just as planets circled in orbits within the solar system.*

There is, however, a fundamental difficulty that had to be faced by all models that involved electrons revolving about a nucleus. A revolving electron undergoes a continual acceleration toward the center and, by Maxwell's electromagnetic theory, such an accelerating charge should be constantly emitting electromagnetic radiation.

Indeed, Nagaoka made that part of his model. The electron in its circular movement about the nucleus acted as a charge oscillating from one end of its orbit to the other, and this should create radiation of corresponding frequency (as in the case of Hertz's spark discharge oscillations, see page 32). If the electron made five hundred trillion revolutions per second in its orbit (which it would do if it traveled at the not-impossible velocity of 150 kilometers per second), it would produce radiation with a frequency of five hundred trillion cycles per second; this would be in the visible light range. Here was an explanation of light as an electromagnetic radiation.

This was so attractive a suggestion that it almost hurts to break it down, but one must. If the revolving electron emits electromagnetic radiation continually, it must lose energy, and kinetic energy (the energy of motion) is all that the electron can lose, as far as we know. Consequently, its motion about the central nucleus must constantly slow and the electron must spiral into the nucleus.†

Since electrons do not, in actual fact, spiral into the nucleus,

---

* This picture of the atom caught the public fancy, perhaps because it compared the atom with something that was already familiar. Although the solar system model was quickly replaced by more complex and more useful models, it has remained in the minds of many nonphysicists. Innumerable science fiction stories, for instance, have been written in which atoms were considered to be tiny solar systems and in which the electron-planets were supposed to be inhabited; sometimes by creatures very much like earthmen.

† As the earth revolves about the sun, it must, by analogy, constantly radiate "gravitational radiation." However, the force of gravity is so much weaker than the electromagnetic force (see page II–164) that the loss of energy by gravi-

another model must be found. Such a model must account not only for the fact that atoms radiate light (and absorb it, too) but that they only radiate and absorb light of certain characteristic wavelengths (or frequencies). By studying the interrelationships of these characteristic wavelengths, hints may be found as to what that structure might be. Hydrogen would be the element to tackle, for it produces the simplest and most orderly spectrum.

Thus, the most prominent line in the hydrogen spectrum has a wavelength of 656.21 millimicrons. Next to that is one at 486.08 millimicrons; then one at 434.01 millimicrons; then one at 410.12 millimicrons; then one at 396.81 millimicrons, and so on. If the wavelengths of these lines are plotted to scale, they will be seen to be separated by shorter and shorter intervals. Apparently, some order exists here.

In 1885, a German mathematician, Johann Jakob Balmer (1825–1898), tinkered with the series of numbers representing the wavelengths of the lines of the hydrogen spectrum and found a simple formula that expressed the wavelength (λ) of the lines. This was:

$$\lambda = \frac{364.56\, m^2}{m^2 - 4} \qquad \text{(Equation 5–1)}$$

where m can have successive whole-number values starting with 3. If $m = 3$, then λ can be calculated as equal to 656.21 millimicrons, which is the wavelength of the first line. If m is set equal to 4, then to 5, and then to 6, the wavelengths of the second, third, and fourth lines of the hydrogen spectrum turn up in the value calculated for λ. This series of lines came to be called the *Balmer series*.

Eventually as m becomes very high, $m^2 - 4$ becomes very little different from m^2, so the two terms would cancel in Equation 5–1. In that case, λ would become equal to 364.56 millimicrons (*Balmer's constant*), and this would be the limit toward which all the lines in the series would tend.

Some years after Balmer's work, the Swedish physicist Johannes Robert Rydberg (1854–1919) put the formula into a more convenient form. He began by taking the reciprocal of both sides of Equation 5–1, and this gave him:

---

tational radiation is insignificant. It would take many trillions of years for the earth to lose a noticeable amount of kinetic energy in this fashion. The electron, subjected to a much stronger force than gravitation, would have its orbit decay in a very short time indeed.

$$\frac{1}{\lambda} = \frac{m^2 - 4}{364.56 \, m^2}$$ (Equation 5–2)

Multiplying both numerator and denominator of the right-hand side of Equation 5–2 by four:

$$\frac{1}{\lambda} = \frac{4(m^2 - 4)}{364.56 \, (4m^2)} = \frac{4}{364.56} \left(\frac{m^2 - 4}{4m^2} \right) = 0.0109 \left(\frac{m^2 - 4}{4m^2} \right)$$
(Equation 5–3)

Let's take each part of the extreme right-hand portion of Equation 5–3 separately. The value 0.0109 is obtained through the division of 4 by Balmer's constant, which is 364.56 millimicrons. The units of the quotient are therefore "per millimicron." Rydberg chose to use the unit "per centimeter." There are 10,000,000 millimicrons in a centimeter, so there are ten million times as many of anything per centimeter as per millimicron. If we multiply 0.0109 by ten million we get 109,000. The exact value, as determined by modern measurements, is 109,737.31 per centimeter. This is called the *Rydberg constant* and is symbolized R. In terms of centimeters, then, we can express Equation 5–3 as follows:

$$\frac{1}{\lambda} = 109,737.31 \left(\frac{m^2 - 4}{4m^2} \right) = R \left(\frac{m^2 - 4}{4m^2} \right) \quad \text{(Equation 5–4)}$$

The value of λ determined by Equation 5–4 is, of course, expressed in centimeters, so that the wavelength of the principal line comes out to 0.000065621 centimeters.

Now consider that portion of the equation which is written $(m^2 - 4)/4m^2$. This can be written as $m^2/4m^2 - 4/4m^2$, or, reducing to lowest terms, $1/4 - 1/m^2$. To make this look more symmetrical, we can now express 4 as a square also and make it $1/2^2 - 1/m^2$. Now Equation 5–4 becomes:

$$\frac{1}{\lambda} = R \left(\frac{1}{2^2} - \frac{1}{m^2} \right)$$ (Equation 5–5)

where m can equal any integer from 3 up.

It is possible to imagine similar series such as:

$$\frac{1}{\lambda} = R \left(\frac{1}{1^2} - \frac{1}{m^2} \right)$$ (Equation 5–6)

$$\frac{1}{\lambda} = R \left(\frac{1}{3^2} - \frac{1}{m^2} \right)$$ (Equation 5–7)

$$\frac{1}{\lambda} = R\left(\frac{1}{4^2} - \frac{1}{m^2}\right) \qquad \text{(Equation 5-8)}$$

and so on. In Equation 5-6, the values of m must be integers greater than 1; in Equation 5-7, integers greater than 3; and in Equation 5-8, integers greater than 4.

The wavelengths given by Equation 5-6 would be shorter than those of the Balmer series and would exist only in the ultraviolet range. This series was actually discovered in 1906 by the American physicist Theodore Lyman (1874–1954) and is consequently known as the *Lyman series*.

The wavelengths given by Equation 5-7 would be longer than those of the Balmer series and would exist only in the infrared range. These were observed in 1908 by the German physicist Friedrich Paschen. The wavelengths given by Equation 5-8 would be still deeper in the infrared, and these were discovered by the American physicist Frederick S. Brackett. Equations 5-7 and 5-8 therefore represent the *Paschen series* and the *Brackett series*, respectively. Other series have also been discovered.

The Bohr Atom

A useful model of the hydrogen atom must therefore not only account for the fact that the circling electron gives off radiation without spiraling into the nucleus, but also that it gives off radiation of highly specific wavelengths, in such a fashion as to make them fit the simple Rydberg equations.

The necessary model was suggested in 1913 by the Danish physicist Niels Bohr (1885–1962). It seemed to him that one ought to apply the then-newly-established quantum theory (see page II–130) to the problem.

If the quantum theory is accepted, then any object which is converting kinetic energy into radiation ought to radiate energy in whole quanta only. This would be true if the earth, for instance, lost energy steadily as it revolved about the sun. The quanta of

Hydrogen spectrum

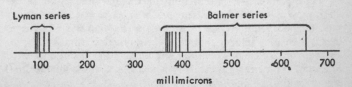

Lyman series   Balmer series

100   200   300   400   500   600   700

millimicrons

energy radiated by the earth in this fashion would, however, be so incredibly small in comparison to the total kinetic energy of the planet that even the most delicate observations would not suffice to detect any unevenness in the motion of the earth. It would seem to be spiraling gradually and smoothly into the sun.

Not so for electrons. The total kinetic energy of so small a body as an electron is not much larger than the individual quanta of visible light. Therefore, if a quantum of visible light is radiated by the electron revolving about the nucleus, a sizable fraction of its kinetic energy is lost all at once. Instead of spiraling gradually inward toward the nucleus (as one would expect according to the tenets of pre-quantum times—that is, of "classical physics"), the electron would suddenly take on a new orbit closer to the nucleus. On the other hand, if light were absorbed by an electron, it would be absorbed only a whole quantum at a time. With the absorption of a whole quantum, an electron would gain a sizable fraction of the energy it already possessed and it would suddenly take on a new orbit farther from the nucleus.

Bohr suggested that the electron had a certain minimum orbit, one that represented its *ground state*; at which time it was as close to the nucleus as it could be, and possessed minimum energy. Such an electron simply could not radiate energy (though the reason for this was not properly explained for over a decade, see page 105). Outside the ground state were a series of possible orbits extending farther and farther from the nucleus. Into these orbits, the *excited states*, the electron could be lifted by the absorption of an appropriate amount of energy.

Bohr arranged the orbits about the nucleus of the hydrogen atom in such a way as to give the electron a series of particular values for its angular momentum. This momentum had to involve Planck's constant (see page II–131) since it was that constant that dictated the size of quanta. Bohr worked out the following equation:

$$p = \frac{nh}{2\pi}$$ (Equation 5–9)

In Equation 5–9, p represents the angular momentum of the electron, h is Planck's constant, and π is, of course, the familiar ratio of the circumference of a circle to its diameter. As for n, that is a positive integer that can take any value from 1 upward. By bringing in Planck's constant and making the electron capable of assuming only certain orbits in which n is a whole number, the atom is said to be *quantized*.

The expression $h/2\pi$ is commonly used in calculations involving the quantized atom, and is usually expressed by the single symbol ℏ, which is referred to as "h bar." Since the value of h is approximately 6.6256×10^{-27} erg-seconds and that of π is approximately 3.14159, the value of ℏ is approximately 1.0545×10^{-27} erg-seconds.

We can therefore express Equation 5–9 as:

$$p = n \, (1.0545 \times 10^{-27}) \qquad \text{(Equation 5–10)}$$

The symbol n is sometimes referred to as a "quantum number" or, more properly, the *principal quantum number*, for there are others. It can be imagined to represent the various orbits. Where n equals 1, it refers to the ground state; where n equals 2, 3, 4, and so on, it refers to the higher and higher excited states.

If the single electron of the hydrogen atom dropped from orbit 2 to orbit 1, it emits a quantum of fixed size, and this is equivalent to a bit of radiation of fixed frequency. This would show up as a bright spectral line in a fixed position. (If the single electron rose from orbit 1 to orbit 2, this would be through the absorption of a quantum of the same fixed size, and this would produce a dark line against a bright background in the same position.)

If the single electron of the hydrogen atom dropped from orbit 3 to orbit 1, this would represent a greater difference in energy, and light of higher frequency would be emitted. Light of still higher frequency would result in a drop of an electron from orbit 4 to orbit 1, and higher frequency still in a drop from orbit 5 to orbit 1.

The series of possible drops from various orbits to orbit 1 would produce a series of successively higher frequencies (or successively lower wavelengths) that would correspond to those in the Lyman series. A series of possible drops from various outer orbits to orbit 2 would give rise to the Balmer series; from various outer orbits to orbit 3 to the Paschen series, and so on.

In the equations defining the wavelengths of the spectral lines included in the various series (Equations 5–5, 5–6, 5–7, and 5–8), the integer in the denominator of the first fraction on the right-hand side of the equation turns out to be the principal quantum number of the orbit into which the electrons drop (or out of which they rise).

If we consider atoms that are more complicated than hydrogen and contain more electrons, we must remember that they also contain nuclei of higher positive charge. The innermost electrons are held progressively more firmly as that nuclear charge increases.

It takes larger increments of energy to move such electrons away from the nucleus into excited states. Conversely, larger quanta of energy are given off when an electron drops closer to its ground state. Whereas the shortest wavelengths hydrogen can produce are those represented by the Lyman series in the ultraviolet, more complicated atoms can produce radiation in the X-ray region. The X-ray wavelength decreases with increasing atomic number, as Moseley had noticed.

So far, so good. If the lines of the hydrogen spectrum had been simple lines, the Bohr model of the hydrogen atom might have been reasonably satisfactory. However, as spectral analysis was refined, it turned out that each line had a *fine structure;* that is, it consisted of a number of distinct lines lying close together. It was as though an electron dropping down to orbit 2, for instance, might drop into any of a number of very closely spaced orbits.

This threatened the quantum interpretation of the atom, but in 1916 the German physicist Arnold Sommerfeld (1868–1951) offered an explanation. Bohr had pictured the electron orbits as

Bohr atom

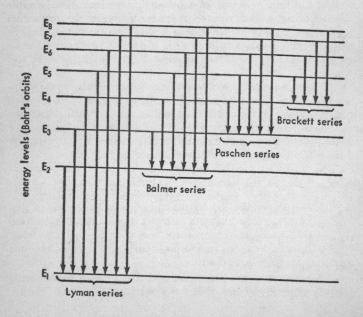

uniformly circular, but Sommerfeld suggested they might also be elliptical. Elliptical orbits of only certain eccentricities could be fitted into the quantum requirements, and for any principal quantum number, a fixed family of orbits—one circular and the rest elliptical—was permissible, the angular momenta among the orbits being slightly different. A drop to each of the various members of the family produced radiation of slightly different frequency.

To take into account the elliptical orbits, Sommerfeld introduced the *orbital quantum number*, for which we can use the symbol L.* The orbital quantum number can have any whole number value from zero up to one less than the value of the principal quantum number. Thus, if $n = 1$, then L can only equal 0; if $n = 2$, then L can equal 0 or 1; if $n = 3$, then L can equal 0, 1, or 2, and so on.

But the spectral lines can be made even more complicated, for in a magnetic field, lines that seem single, split further. In order to account for this, a third number, the *magnetic quantum number* had to be introduced, and this was symbolized as m.

The magnetic quantum number was visualized as extending the family of orbits through three-dimensional space. Not only could an orbit be elliptical rather than circular, but it could also be tilted to the principal orbit by varying amounts. The possible values of m are the same as those for L, except that negative values are also included. Thus if $n = 2$—so that L could be either 0 or 1—m could be either 0, 1, or -1. If $n = 3$—so that L could be either 0, 1, or 2—m could be either 0, 1, 2, -1, or -2, and so on.

Finally, a fourth and last quantum number had to be introduced, the *spin quantum number*, symbolized as s. This was visualized as representing the spin of the electron (analogous to the rotation of the earth about its axis). This spin could either be clockwise or counterclockwise, so that in connection with any value of n there can be only two values of s, $1/2$ and $-1/2$.

Sub-shells

The orbits described by the quantum numbers are all that are available. When an atom contains more than one electron (as is true for all atoms other than those of hydrogen), they must be distributed among these orbits, filling them from the one closest to the nucleus outward.

* Actually, the usual symbol is the lower-case "l," which I am not using here because of the ease with which it is confused with the numeral "1".

But how many electrons may be permitted in each orbit? In 1925, the Austrian physicist Wolfgang Pauli (1900–1958) suggested that in order to account for the various spectral characteristics of the different elements, one must assume that no two electrons in a given atom can have all four quantum numbers identical. This means that in any orbit (circular, elliptical, or tilted) two electrons at most may be present; and of these two, one must spin clockwise and the other must spin counterclockwise. Thus, the presence of two electrons of opposite spin in a given orbit excludes other electrons, and this is called Pauli's *exclusion principle*.

We can now determine the number of electrons that can be included in each of the different orbit-families represented by the principal quantum numbers.

Suppose $n = 1$. In that case, $L = 0$ and $m = 0$. No other combinations are possible, and the only orbit is 1/0/0. This may contain two electrons of opposite spins. The total number of electrons which may be contained in the first orbit-family ($n = 1$) is therefore 2.

Next, suppose that $n = 2$. In that case, L can equal either 0 or 1, and m can equal 0, 1, or -1. To be more specific, if $n = 2$ and $L = 0$, then m must equal 0, too. If $n = 2$ and $L = 1$, then m can equal either 0, 1, or -1. There are therefore four possible orbits for $n = 2$. These are 2/0/0, 2/1/0, 2/1/1, and 2/1/-1. In each one of these orbits two electrons of opposite spins can exist; consequently, the second orbit-family ($n = 2$) can contain eight electrons altogether.

By similar reasoning, the total number of electrons that can be present in the next orbit-family ($n = 3$) turns out to be eighteen. In fact, it can be shown that the maximum number of electrons in any orbit-family is equal to $2n^2$. Therefore for orbit-family $n = 4$, a total of thirty-two electrons may be found; for orbit-family $n = 5$, a total of fifty electrons may be found, and so on.

The orbit-families, represented by the principal quantum number n and deduced from physical data, correspond to the different electron shells deduced from chemical data and made use of in the Lewis-Langmuir model of the atom.

The number of electrons in each orbit-family or electron shell can be divided into *sub-shells*, according to the value of L. Thus, where $n = 1$, L can equal only 0, so that the first electron shell consists of only a single sub-shell, which can hold two electrons.

Where $n = 2$, on the other hand, L can equal either 0 or 1. Where $L = 0$, only one orbit (2/0/0), containing two electrons at most, is possible; but where $L = 1$, three orbits (2/1/0, 2/1/1, and 2/1/−1) containing, at most, a total of six electrons are possible. The eight electrons of the second shell can, therefore, be divided into two sub-shells, one of 2 electrons and one of 6.

In the same way, it can be shown that the eighteen electrons of the third electron shell can be divided into three sub-shells, one capable of holding 2 electrons, one of holding 6, and one of holding 10. In general, the electron shell of principal quantum number n can be divided into n sub-shells, where the first can contain two electrons, and where each one thereafter can contain four more than the one before (6, 10, 14, 18, and so on).

The subgroups are symbolized as s, p, d, f, g, h, and i. We may therefore say that the first electron shell contains only the $1s$ sub-shell, that the second electron shell contains a $2s$ sub-shell and a $2p$ sub-shell, and so on.

Now let's see how all this applies to the individual elements. The first two present no problem. Hydrogen has one electron and helium two, and both can be accommodated in the single sub-shell of the first electron shell.

	Number of Electrons in $1s$	Atomic Number
Hydrogen	1	1
Helium	2	2

All elements of atomic number greater than 2 contain two electrons in the first shell and distribute the remainder among the outer shells. The elements immediately after helium make use of the second electron shell, which is made up of sub-shells $2s$ (electron capacity, two) and $2p$ (electron capacity, six).

	Number of Electrons in			Atomic Number
	$1s$	$2s$	$2p$	
Lithium	2	1	—	3
Beryllium	2	2	—	4
Boron	2	2	1	5
Carbon	2	2	2	6
Nitrogen	2	2	3	7
Oxygen	2	2	4	8
Fluorine	2	2	5	9
Neon	2	2	6	10

In neon the second electron shell is full, and elements of higher atomic number must place electrons in the third electron shell. The third shell has three sub-shells, $3s$, $3p$ and $3d$ with electron capacities of two, six and ten, respectively.

| | \multicolumn{5}{c}{Number of Electrons in} | Atomic Number |
	$1s$	$2s$ $2p$	$3s$	$3p$	$3d$	
Sodium	2	8	1	—	—	11
Magnesium	2	8	2	—	—	12
Aluminum	2	8	2	1	—	13
Silicon	2	8	2	2	—	14
Phosphorus	2	8	2	3	—	15
Sulfur	2	8	2	4	—	16
Chlorine	2	8	2	5	—	17
Argon	2	8	2	6	—	18

Notice the similarities in the electron configurations of these atoms and the ones in the previous list. The $2s/2p$ configuration of lithium is like the $3s/3p$ configuration of sodium. There is the same comparison to be made between beryllium and magnesium, between boron and aluminum, and so on. No wonder the periodic table is as it is.

Argon, which has a $3s/3p$ combination of 2/6, just as neon has a $2s/2p$ combination of 2/6, is also an inert gas. Yet argon does not have its outermost shell completely filled. There is still room for ten more electrons in the $3d$ sub-shell. The conclusion must be that the inert gas properties are brought on not by a truly completed electron shell but only by the complete filling of the s and p sub-shells. These two sub-shells always contain a total of eight electrons; consequently, this total of eight in the outermost shell is the hallmark of the inert gas. The one exception is helium. That contains electrons only in the first electron shell, which is made up of the $1s$ sub-shell only. It therefore contains only two electrons in its outermost (and only) electron shell.

You might suppose that the elements immediately after argon would possess electrons in sub-shell $3d$. This is not so, however. We can understand this by viewing the sub-shells of a particular electron shell as taking up room. As one goes outward from the nucleus, each succeeding electron shell has more sub-shells, and eventually the outer sub-shells of one will begin to overlap the inner sub-shells of the next one out. In this case, $3d$, the outermost sub-shell of $n = 3$ overlaps $4s$, the innermost sub-shell of $n = 4$ and it is $4s$ that is therefore next to be filled. Thus:

	Number of Electrons in					Atomic Number
	$1s$	$2s$ $2p$	$3s$ $3p$	$3d$	$4s$	
Potassium	2	8	8	—	1	19
Calcium	2	8	8	—	2	20

Potassium, with a single electron in sub-shell $4s$, is like sodium with a single electron in sub-shell $3s$ and lithium, with a single electron in sub-shell $2s$. Calcium similarly resembles magnesium and beryllium.

Transition Elements

If scandium, the element after calcium, possessed an electron in sub-shell $4p$, it would have an s/p combination of 2/1, and would resemble aluminum in its properties. However, this is not what happens. With sub-shell $4s$ filled in the case of calcium, the additional electrons of the next few elements are added to sub-shell $3d$ as follows:

	Number of Electrons in					Atomic Number
	$1s$	$2s$ $2p$	$3s$ $3p$	$3d$	$4s$	
Scandium	2	8	8	1	2	21
Titanium	2	8	8	2	2	22
Vanadium	2	8	8	3	2	23
Chromium	2	8	8	5	1	24
Manganese	2	8	8	5	2	25
Iron	2	8	8	6	2	26
Cobalt	2	8	8	7	2	27
Nickel	2	8	8	8	2	28
Copper	2	8	8	10	1	29
Zinc	2	8	8	10	2	30

The overlapping of sub-shells $3d$ and $4s$ is not very pronounced, so there is not much difference between a $3d/4s$ arrangement of 5/1 and 4/2, or of 10/1 and 9/2. In the case of chromium and copper there are reasons for preferring to assign only one electron to $4s$, but this is a mere detail.

What is important in the ten elements from scandium to zinc inclusive is that the difference in electron arrangement concentrates on an inner sub-shell, 3*d*. The outermost sub-shell, 4*s*, is the same (or virtually the same), in all. This series of elements makes up a group of *transition elements,* and the progressive difference among them in chemical properties is not as sharp as among the succession of elements from hydrogen to calcium, where it is the outermost sub-shell in which the difference in electron distribution shows up.

Indeed, the three successive elements—iron, cobalt, and nickel—resemble each other so closely as to form a tight-knit family group of elements.

The Lewis-Langmuir model of the atom makes no allowance for changes in the electron content of inner shells, and it is for this reason that this model does not work well for transition elements (and, as it happens, about three-fifths of all the elements are transition elements).

With zinc, the third electron shell is completely filled and contains a grand total of eighteen electrons. Sub-shell 4*s* is also filled and additional electrons must be added to 4*p* and beyond. Sub-shell 4*p* has a capacity of six electrons:

| | Number of Electrons in | | | | | Atomic Number |
	1*s*	2*s* 2*p*	3*s* 3*p* 3*d*	4*s*	4*p*	
Gallium	2	8	18	2	1	31
Germanium	2	8	18	2	2	32
Arsenic	2	8	18	2	3	33
Selenium	2	8	18	2	4	34
Bromine	2	8	18	2	5	35
Krypton	2	8	18	2	6	36

These six elements have the same *s/p* electron configuration as do the series of elements from aluminum to argon or from boron to neon. Thus gallium resembles aluminum and boron in its properties; germanium resembles carbon and silicon, and so on. Krypton with a 4*s*/4*p* combination of 2/6 is, of course, an inert gas.

The fourth shell has two additional sub-shells: 4*d,* which can hold ten electrons, and 4*f,* which can hold fourteen. Both these sub-shells overlap the innermost sub-shell of the fifth shell, 5*s*:

	Number of Electrons in							Atomic Number
	1s	2s 2p	3s 3p 3d	4s 4p	4d	4f	5s	
Rubidium	2	8	18	8	—	—	1	37
Strontium	2	8	18	8	—	—	2	38

The next elements possess electrons in sub-shell $4d$ so that a new series of transition elements is produced like those from scandium to zinc inclusive. Thus, we have:

	Number of Electrons in							Atomic Number
	1s	2s 2p	3s 3p 3d	4s 4p	4d	4f	5s	
Yttrium	2	8	18	8	1	—	2	39
Zirconium	2	8	18	8	2	—	2	40
Niobium	2	8	18	8	4	—	1	41
Molybdenum	2	8	18	8	5	—	1	42
Technetium	2	8	18	8	5	—	2	43
Ruthenium	2	8	18	8	7	—	1	44
Rhodium	2	8	18	8	8	—	1	45
Palladium	2	8	18	8	10	—	—	46
Silver	2	8	18	8	10	—	1	47
Cadmium	2	8	18	8	10	—	2	48

We next return to the $5p$ column and produce half a dozen elements with the s/p combination similar to that of the series of elements from boron to neon. These are not transition elements.

	Number of Electrons in							Atomic Number
	1s	2s 2p	3s 3p 3d	4s 4p 4d	4f	5s	5p	
Indium	2	8	18	18	—	2	1	49
Tin	2	8	18	18	—	2	2	50
Antimony	2	8	18	18	—	2	3	51
Tellurium	2	8	18	18	—	2	4	52
Iodine	2	8	18	18	—	2	5	53
Xenon	2	8	18	18	—	2	6	54

Xenon is another inert gas.

There remains sub-shell 4f, capable of holding fourteen electrons. There are also sub-shells 5d, 5f and 5g capable of holding 10, 14, and 18 electrons, respectively. All of these overlap sub-shell 6s, however.

	Number of Electrons in										Atomic Number
	1s	2s	3s	4s	4f	5s	5d	5f	5g	6s	
		2p	3p	4p		5p					
			3d	4d							
Cesium	2	8	18	18	—	8	—	—	—	1	55
Barium	2	8	18	18	—	8	—	—	—	2	56

With lanthanum, the element beyond barium, electrons start entering sub-shell 4f or 5d, and this gives us a new sort of transition element. In the ordinary transition elements from scandium to zinc or from yttrium to cadmium, the sub-shell to which electrons were being added was covered only by the one or two electrons in the next higher s sub-shell. Here, however, where 4f is involved, the electrons being added are covered not only by two electrons in 5s but by six electrons in 5p and by two electrons in 6s. In these elements, electrons are being added to a sub-shell that is deeper within the atom, so to speak, than was true in the case of the transition elements considered earlier. The sub-shell in which the electron difference occurs is more efficiently covered by outer electrons. For this reason, these elements (the lanthanides) resemble each other particularly closely.

	Number of Electrons in										Atomic Number
	1s	2s	3s	4s	4f	5s	5d	5f	5g	6s	
		2p	3p	4p		5p					
			3d	4d							
Lanthanum	2	8	18	18	—	8	1	—	—	2	57
Cerium	2	8	18	18	1	8	1	—	—	2	58
Praseodym- ium	2	8	18	18	3	8	—	—	—	2	59
Neodymium	2	8	18	18	4	8	—	—	—	2	60
Promethium	2	8	18	18	5	8	—	—	—	2	61
Samarium	2	8	18	18	6	8	—	—	—	2	62
Europium	2	8	18	18	7	8	—	—	—	2	63
Gadolinium	2	8	18	18	7	8	1	—	—	2	64
Terbium	2	8	18	18	8	8	1	—	—	2	65

	1s	2s 2p	3s 3p 3d	4s 4p 4d 4f	5s 5p	5d	5f	5g	6s	
Dysprosium	2	8	18	18	9	8	1	—	—	2	66
Holmium	2	8	18	18	10	8	1	—	—	2	67
Erbium	2	8	18	18	11	8	1	—	—	2	68
Thulium	2	8	18	18	13	8	—	—	—	2	69
Ytterbium	2	8	18	18	14	8	—	—	—	2	70
Lutetium	2	8	18	18	14	8	1	—	—	2	71

The elements after lutetium add further electrons to sub-shell 5d, which can hold up to ten electrons. This sub-shell is still covered by the two electrons in sub-shell 6s, so that we continue after lutetium with a set of ordinary transition elements:

	1s	2s 2p	3s 3p 3d 4f	4s 4p 4d	5s 5p	5d	5f	5g	6s	Atomic Number
					Number of Electrons in					**Atomic**
Hafnium	2	8	18	32	8	2	—	—	2	72
Tantalum	2	8	18	32	8	3	—	—	2	73
Tungsten	2	8	18	32	8	4	—	—	2	74
Rhenium	2	8	18	32	8	5	—	—	2	75
Osmium	2	8	18	32	8	6	—	—	2	76
Iridium	2	8	18	32	8	7	—	—	2	77
Platinum	2	8	18	32	8	9	—	—	1	78
Gold	2	8	18	32	8	10	—	—	1	79
Mercury	2	8	18	32	8	10	—	—	2	80

With mercury, sub-shell 5d is filled. Sub-shells 5f and 5g remain untouched, and electrons are next found in sub-shell 6p, so that we have a group of elements with the familiar s/p arrangement of the boron-to-neon group:

	1s	2s 2p	3s 3p 3d	4s 4p 4d 4f	5s 5p 5d	5f	5g	6s	6p	Atomic Number
Thallium	2	8	18	32	18	—	—	2	1	81
Lead	2	8	18	32	18	—	—	2	2	82
Bismuth	2	8	18	32	18	—	—	2	3	83
Polonium	2	8	18	32	18	—	—	2	4	84
Astatine	2	8	18	32	18	—	—	2	5	85
Radon	2	8	18	32	18	—	—	2	6	86

Radon is an inert gas.

There still remain sub-shells 5*f* and 5*g*, with capacities for 14 and 18 electrons, respectively. There are also sub-shells 6*d*, 6*f*, 6*g* and 6*h*, with electron capacities of 10, 14, 18, and 22 respectively. All of these, however, overlap sub-shell 7*s*.

					Number of Electrons in								Atomic
1s	2s	3s	4s	5s 5f 5g			6s	6d	6f	6g	6h	7s	Number
	2p	3p	4p	5p		6p							
		3d	4d	5d									
			4f										
Francium	2	8	18	32	18	—	—	8	—	—	—	— 1	87
Radium	2	8	18	32	18	—	—	8	—	—	—	— 2	88

There then arises just such a situation as occurs in the case of the lanthanides:

					Number of Electrons in							Atomic Number
	1s	2s	3s	4s	5s	5f 5g 6s	6d	6f	7s	
		2p	3p	4p	5p	6p		6g		
			3d	4d	5d			6h		
				4f						
Actinium	2	8	18	32	18	— — 8	1	—	2	89
Thorium	2	8	18	32	18	— — 8	2	—	2	90
Protactinium	2	8	18	32	18	2 — 8	1	—	2	91
Uranium	2	8	18	32	18	3 — 8	1	—	2	92
Neptunium	2	8	18	32	18	4 — 8	1	—	2	93
Plutonium	2	8	18	32	18	5 — 8	1	—	2	94
Americium	2	8	18	32	18	7 — 8	—	—	2	95
Curium	2	8	18	32	18	7 — 8	1	—	2	96
Berkelium	2	8	18	32	18	8 — 8	1	—	2	97
Californium	2	8	18	32	18	9 — 8	1	—	2	98
Einsteinium	2	8	18	32	18	10 — 8	1	—	2	99
Fermium	2	8	18	32	18	11 — 8	1	—	2	100
Mendelevium	2	8	18	32	18	12 — 8	1	—	2	101
Nobelium	2	8	18	32	18	13 — 8	1	—	2	102
Lawrencium	2	8	18	32	18	14 — 8	1	—	2	103

The group of elements from actinium to lawrencium are the actinides. When the element with atomic number 104 is studied, it is fully expected that the 104th electron will be added to sub-shell $6d$ and that this element will resemble hafnium in its chemical properties.

If you will now compare the electron arrangements of the various elements as given in this section with the periodic table presented on page 16, you will see how the periodic table reflects similarities in electron arrangements.

6

Electron Energy Levels

Semiconductors

While the notion of electron shells and sub-shells finally rationalized the periodic table, even down to the until-then puzzling lanthanides, the Bohr model itself, even as modified by Sommerfeld and others, did not stand up in its original form. The attempt to produce a literal picture of the atom as consisting of electron particles moving in orbits that were circular, elliptical and tilted—much more complicated than the solar system, but still with some key points of similarity—grew top-heavy and collapsed.

During the early 1920's, it became more common to think, not of orbits, but of *energy levels*. Electrons moved from one energy level to another and the difference in energy levels determined the size of the quantum (hence the frequency of the radiation) emitted or absorbed.

In 1925, in fact, the German physicist Werner Heisenberg (1901–   ) worked out a system whereby the energy levels of atoms could be written out as a set of numbers. These could be arranged in rectangular arrays called "matrices," and these matrices could be manipulated according to mathematical principles (*matrix algebra*). The proper manipulation applied to atomic data (*matrix mechanics*) produced values from which spectral lines could be calculated. No actual picture of any sort was required for the atom by this view; it had faded away completely into a mere collection of numbers.

In the case of a single atom, energy levels could be pictured as simple lines at given heights above the base of a schematic drawing. Two electrons of opposite spin could occupy any of the energy levels and could shift from one level to any other that was not fully occupied. The spaces between the lines represented "forbidden gaps" within which no electron could be located. Each element had its own characteristic collection of lines and gaps, of course.

If two atoms of an element are in close proximity, the picture becomes more complicated. The outer electrons of the two atoms are close enough for the energy levels to merge. For each energy level, the electron population is doubled. An energy level cannot hold more than its capacity (two electrons of opposite spin); consequently, what happens is that the energy levels associated with the two atoms shift a bit—one becoming slightly higher than the other. Each can then hold its own electrons.

In a solid, where there are a vast number of atoms existing in near proximity, this happens on a grand scale. An energy level can no longer be depicted as a line but as a dense assemblage of lines at slightly different heights. What was an energy line is now actually an *energy band*. Electrons can rise from band to band, rather than from line to line, and there are forbidden gaps between the bands.

If each of the atoms making up a solid has its outermost energy levels containing all the electrons they can hold, then the result is a filled energy band. In such a case, the electrons are fixed in place. They cannot pass from one atom to the next since the neighboring atom has no room for it. Such a solid is a nonconductor of electricity. Extreme examples are sulfur and quartz.

If the outermost energy levels of the individual atoms contain fewer electrons than capacity, the resulting energy band of the solid is only partly filled, and electrons can move easily from atom to atom by way of unfilled energy levels. The electric impulse can easily travel across a path containing these "free electrons," and the substance is a conductor of electricity. Extreme examples are silver and copper.

Even if the energy band is electron-filled, there is a chance of electrical conduction. Above the filled energy band is another energy band that is empty. The absorption of energy may kick a few electrons up into the higher energy band and there they may move freely. The likelihood of this happening depends on the width of the forbidden gap between the filled energy band and the higher empty one. If the gap is quite wide, the likelihood of an electron leaping across is low, and the substance is an excellent nonconductor.

In some substances (for example, the elements silicon and germanium) the forbidden gap is comparatively narrow, and the likelihood of an electron leap into the higher band becomes appreciable. The result is a *semiconductor*. If the temperature is raised, the tendency of an electron to reach the higher band is increased, for more energy becomes available to kick it upward. For this reason, the resistance of a semiconductor decreases with temperature. (A semiconductor differs in this respect from a metallic conductor, on which the chief effect of heightened temperature is to produce intensified atomic vibrations that interfere with transmission of the electrical impulse and increase the resistance.)

The semiconductors have proven unexpectedly and fabulously useful, where their chemical composition and physical structure are suitably tailored to need.

Consider germanium, for instance. Like carbon, the germanium atom has four electrons in its outermost shell. Each germanium atom can contribute one electron to form a shared pool of two with each of four other germanium atoms. In the end, the germanium atoms will be stacked in such a fashion that each is connected with four others. Under such conditions, all electrons are firmly in place and the substance's semiconducting properties are at a minimum.

In order for this to happen, all the atoms must be stacked perfectly. Imperfections in the crystal means that some atoms are going to be out of place with respect to their neighbors and will not be able to share electrons. It is these few unshared electrons that contribute to the semiconducting properties of germanium.

Those properties are more useful if they arise out of a deliberately added impurity, rather than out of the random imperfections that are almost inevitably found in any germanium crystal. Imagine a perfect germanium crystal formed out of germanium to which a small trace of arsenic had been added as an impurity. Arsenic has five electrons in its outermost shell. When an arsenic atom tries to fit in with the germanium arrangement, it can find room for four of its electrons in the shared pools formed with neighboring germanium atoms. The fifth arsenic electron, however, is at loose ends. It acts as a free electron.

Under the influence of an electric potential applied across

the crystal, the free electrons, negatively charged of course, drift away from the negative electrode and toward the positive. Because it is a negatively-charged particle that is drifting, the result is an *n-type semiconductor*, "n" for negative.

Next, consider a germanium crystal to which a trace of boron has been added. The boron atom has three electrons in its outermost shell. Each of the three can join in a shared pool with electrons of a neighboring germanium atom. But only three germanium atoms can be thus accommodated; the fourth will be left with a "hole" where an electron ought to be.

Under the influence of an electric potential across such a crystal, a negatively-charged electron will be pushed or pulled into the hole, traveling always from the side of the repelling negative electrode toward the attracting positive electrode. But the electron that has filled the hole has left another hole in the place it had earlier occupied. Since the electron came from the direction of the negative electrode, the new hole is now closer to the negative electrode than the old hole had been. The same thing happens over and over again, and the hole drifts steadily toward the negative electrode and away from the positive one. Indeed, the drifting hole behaves as though it were a positively-charged particle, and so such a crystal is a *p-type semiconductor*, "p" for positive.

Solid-State Devices

The drift of electrons through a semiconductor can be governed and manipulated in order to achieve the ends that had earlier been achieved by a vacuum tube. The key difference is that electrons move across solids in the former case and across a vacuum in the latter. For this reason, electronic instruments in which semiconductors play a part are called *solid-state devices*.

Electrons and holes

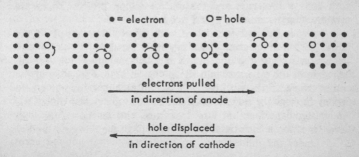

● = electron   O = hole

electrons pulled
in direction of anode

hole displaced
in direction of cathode

Consider, for instance, a crystal of which one half is n-type and the other half is p-type. Imagine that the n-type end is connected to the negative pole of a battery, while the p-type is attached to the positive pole. If the circuit is closed, the electrons in the n-type end of the crystal are repelled from the negative pole and move toward the junction between the two halves. Meanwhile the holes in the p-type end are repelled from the positive pole and also move toward the junction. There they can meet and neutralize each other, while electrons flood into the n-type end and are withdrawn from the p-type end (making new holes). By the movement of electrons and holes, ever-renewed as long as the circuit is closed, current can flow across the crystal.

But suppose the circuit is arranged in the opposite way, with the n-type end of the crystal attached to the positive pole of the battery and the p-type attached to the negative pole. Now the electrons in the n-type end are attracted to the positive pole and move away from the junction. The holes in the p-type end are attracted to the negative pole and also move away from the junction. First the junction area and then the entire crystal is emptied of both free electrons and of holes; consequently, it becomes a nonconductor and no current flows.

In short, such an n-p crystal acts to allow current to pass in only one direction. If connected to a source of alternating current, it would serve to rectify the current. In fact the n-type end of the crystal is similar in its behavior to the heated filament of a vacuum tube, while the p-type end is similar in its behavior to the plate. The crystal is like a diode with the two parts meeting to form a junction. Such a device is therefore called a *junction diode*.

It is possible to construct a semiconductor analog of a triode also. In an ordinary triode, a third element, the grid, is inserted between the filament and the plate, and the same may be done in solid-state devices. A crystal can be made up of three regions, with both ends n-type and the middle region p-type. There are now two junctions, an n-p and a p-n.

Imagine one n-type end of such a crystal attached to the negative pole of a battery, and the other attached to the positive pole. Electrons at the end attached to the negative pole are repelled from the pole and into the p-type middle. Electrons at the end attached to the positive pole are attracted toward the pole and away from the middle, so that electrons are pulled out of the p-type middle. Electrons flow from one end to the other, with the p-type middle encouraging the first half of the flow and hindering the second half. The rate of flow can be sharply altered, then, by the size of the charge placed on the p-type center.

Such a "junction triode" was first made practical in 1948 by the English-American physicist William Bradford Shockley (1910–) and his American co-workers John Bardeen (1908–) and Walter House Brattain (1902–). Because the device transferred a current across a material that ordinarily had high resistance (a "resistor"), it was called by a shortened version of the phrase "transfer-resistor." It was a *transistor*.

In the early days of radio, before the vacuum tube had been devised, natural crystals had sometimes been found that in proper combination with other materials displayed rectifying action. It was the use of these that gave the name to the old-fashioned "crystal sets."

The development of the vacuum tube had thrown the crystals out of use, but now the specially tailored transistor crystals turned the tables. Transistors had several advantages over the vacuum tubes. They required no vacuum but were solid throughout; as a result they were sturdier than tubes. They made no use of heat (as was required in a vacuum tube, for there only a hot filament would emit electrons), so they required no warm-up period and also lasted longer. Furthermore, they could be made much, much smaller than vacuum tubes.

Since it was the vacuum tube that made the radio so bulky, the use of transistors made it possible to develop radios no larger than a pack of cigarettes. In fact, "to transistorize" came to be synonymous with "to miniaturize." The effect of transistorization was most spectacular in connection with electronic computers requiring thousands of vacuum tubes. The size of computers was drastically reduced with the coming of solid-state devices.

Semiconductors could also be used in the direct conversion of heat into electricity (*thermoelectricity*). The general phenomenon whereby a temperature difference can be made to give rise to an electric current was first observed in 1821 by the German physicist Thomas Johann Seebeck (1770–1831). He found that when part of a circuit made up of two different metals is heated, a magnetic needle would be deflected if placed near the point at which the two metals meet. This is the *Seebeck effect*.

The effect on the magnet came from the magnetic field set up by the electric current that was produced by the appearance of a temperature difference. Unfortunately, Seebeck did not interpret his observations in this fashion (the connection between electricity and magnetism was just being discovered then—see page II–201). He thought it purely a magnetic effect. Consequently, interest in the effect languished, and it is only in recent decades that it has revived.

Consider an n-type semiconductor, one end of which is heated. The hot end is more likely to have its electrons kicked into a higher energy band, wherein they can easily drift away. As a result, electrons tend to move from the hot end toward the cold end, and an electric potential (cold end, negative; hot end, positive) is set up. (This would also happen in an ordinary conductor, but an ordinary conductor has a high concentration of free electrons at the cold end, which would repel incoming electrons from the hot end; consequently, there would only be a small net drift. In a semiconductor, the number of free electrons at the cold end is very small, and the repelling effect is much weaker. For that reason a much larger drift takes place, and a much higher potential difference is set up in a semiconductor than in an ordinary conductor.)

If a p-type semiconductor is used, the holes are more easily filled at the hot end, where the electrons are made more mobile by the greater energy associated with high temperature. New holes are formed further away from the hot end and, in brief, the holes drift from hot end to cold. Again a potential difference is set up but this time with the cold end positive and the hot end negative.

If an n-type and a p-type semiconductor are joined at the hot end, electrons will flow from the p-type cold end to the hot junction and then on to the n-type cold end. If the two ends are connected through a closed circuit, a current will flow and useful work will be done as long as the temperature difference is maintained. A kerosene lamp, then, can be used to power an electric generator without moving parts.

The situation can be reversed. If an electric current is forced through a circuit made up of different materials, a temperature difference is set up. This was first observed in 1834 by a French physicist, Jean Charles Athanase Peltier (1785–1845), and is therefore called the *Peltier effect*. By using p-type and n-type semiconductors joined at one end, the heat of the system can be concentrated at that end and removed from the other, so that the device can function either as a heater or a refrigerator.

Semiconductors can also convert light to electricity. Essentially, a device of this sort, a *solar battery*, consists of an n-type semiconductor overlaid with a thin layer of p-type semiconductor. There is a potential difference between the electron-rich n-type section and the electron-poor p-type section which could set up a current for a very short time until the electrons had flooded into the p-type area and filled the holes.

If the plate is exposed to sunlight, the high-energy quanta of the light act to knock electrons loose in the p-type area, forming more holes as fast as incoming electrons fill them and thus allowing a continuous current of significant proportions to flow through a circuit connected to the battery. Such batteries will hold good for extended periods of time, and some have successfully been used to power artificial satellites for years.

Masers and Lasers

The fact that energy levels are separated by fixed distances and that a change of levels can only be accomplished by the absorption or emission of photons of specific size has given rise in recent years to important new devices.

Consider the ammonia molecule (NH_3), for instance. It possesses two energy levels separated by a gap that is equal in size to the energy content of a photon equivalent to an electromagnetic wave with a frequency of 24 billion cycles per second. The wavelength of such radiation is 1.25 centimeters, which places it in the microwave portion of the electromagnetic spectrum.

This difference in energy levels can be pictured in terms of the architecture of the molecule. The three hydrogen atoms of the ammonia molecule can be viewed as occupying the three vertices of an equilateral triangle, while the single nitrogen atom is some distance above the center of the triangle. With the change in energy level, the nitrogen atom moves through the plane of the triangle to an equivalent position on the other side. The ammonia molecule can be made to vibrate back and forth with a frequency of just 24 billion times a second.

This vibration period is extremely constant, much more so than the period of any man-made vibrating device; much more constant, even, than the movement of astronomical bodies. A vibrating molecule, producing a microwave of highly constant frequency, can therefore be used to control time measuring devices with unprecedented precision. By means of such *atomic clocks*, accuracies in time measurement of one second in 100,000 years are looked forward to as attainable.

But let us leave molecular architecture and consider only energy levels. If a beam of microwaves passes through ammonia gas, a beam containing photons of the proper size, the ammonia molecules will be raised to the higher energy level. In other words, a portion of the microwave beam will be absorbed.

What if an ammonia molecule is already in the higher energy level, however? As early as 1917, Einstein pointed out that if a photon of just the right size struck such an upper-level molecule, the molecule would be nudged back down to the lower level and would emit a photon of exactly the size—and moving in exactly the direction—of the entering photon.

Ammonia exposed to microwave radiation could, therefore, undergo two possible changes: Molecules could be pumped up from lower level to higher, or be nudged down from higher level to lower. Under ordinary conditions, the former process would predominate, for only a very small percentage of the ammonia molecules would at any one instant be at the higher energy level.

Suppose, though, that some method were found to place all or almost all the molecules in the upper energy level. Then it would be the movement from higher level to lower that would predominate. Indeed, something quite interesting would happen. The incoming beam of microwave radiation would supply a photon that would nudge one molecule downward. A second photon would be released, and the two would speed on, striking two molecules, so that two more were released. All four would bring about the release of four more, and so on. The initial photon would let loose a whole avalanche of photons, all of exactly the same size and moving in exactly the same direction.

Physicists both in the United States and the Soviet Union labored to achieve such a situation, but the lion's share of credit for success goes to the American physicist Charles Hard Townes (1915–   ). In 1953, he devised a method for isolating excited ammonia molecules and subjecting them to stimulation by microwave photons of the exact energy content. A few photons entered, and a flood of photons left. The incoming radiation was thus greatly amplified.

The process was described as "microwave amplification by stimulated emission of radiation." The phrase was initialed as "m. a. s. e. r." and the instrument came to be known by the acronym *maser*, a word which quickly replaced the more dramatic, but too narrow, phrase "atomic clock."

Solid-state masers were soon developed, using paramagnetic atoms or molecules (see page II–155) in a magnetic field. Here an electron can be pictured as occupying two energy levels, depending upon its spin: the lower level, when it is spinning in a direction parallel to the magnetic field; the upper, when it is spinning in the opposite direction. The electrons are slowly pumped upward to the higher level and then made to release all their stored energy in a sudden burst of radiation at a single frequency (*monochromatic radiation*).

The first masers, both gaseous and solid state, were intermittent. That is, they had to be pumped up first and then released. After a burst of radiation, no more could be emitted until the pumping progress had been repeated.

To get round this, it occurred to the Dutch-American physicist Nicholaas Bloembergen (1920–   ) to make use of a three-level system. This is possible, for instance, if the solid that forms the radiating core of the maser contains atoms of metals such as chromium or iron. It then becomes possible to distribute electrons among three energy levels, a lower, a middle and an upper. In this case, both pumping and stimulated emission can go on simultaneously. Electrons are pumped up from the lowest energy level to the highest. Once at the highest, proper stimulation will cause them to drop down first to the middle level, then to the lower. Photons of different size are required for pumping and for stimulated emission, and the two processes will not interfere. Thus, we end with a continuous maser.

As microwave amplifiers, masers can be used as very sensitive detectors in radio astronomy, where exceedingly feeble microwave beams received from outer space will be greatly intensified with great fidelity to the original radiation characteristics. (Reproduction without loss of original characteristics means to reproduce with little "noise." Masers are extraordinarily "noiseless" in this meaning of the word.)

In principle, the master technique could be applied to electromagnetic waves of any wavelength, notably to those of visible light. Townes pointed out the possibility of such applications to light wavelengths in 1958. Such a light-producing maser might be called an *optical maser*. Or this particular process might be called "light amplification by stimulated emission of radiation," and the acronym *laser* might be used. It is the latter that has grown popular.

The first successful laser was constructed in 1960 by the American physicist Theodore Harold Maiman (1927–   ). He used a bar of synthetic ruby for the purpose—this being, essentially, aluminum oxide with a bit of chromium oxide added. (It is the chromium oxide that lends the synthetic ruby its red color.) If the ruby bar is exposed to light, the electrons of the chromium atoms are pumped to higher levels and after a short while begin to fall back. The first few photons of light emitted (with a wavelength of 694.3 millimicrons) stimulate the production of other such photons, and the bar suddenly emits a beam of deep red light. Before the end of 1960, continuous lasers were prepared.

The laser made possible light in a completely new form. The light was the most intense and the most narrowly monochromatic that had ever been produced, but it was even more than that.

Light produced in any other fashion, from a wood fire to the sun, consists of relatively short wave packets oriented in all conceivable directions. Ordinary light is made up of countless numbers of these packets.

The light produced by a stimulated laser, however, consists of photons of the same size and moving in the same direction. This means that the wave packets are all of the same frequency, and since they are lined up precisely end to end, so to speak, they melt together. This is *coherent light*, because the wave packets seem to stick together. Physicists had learned to prepare coherent radiation for radiation of long wavelength, like radio waves. (It is a coherent radio wave that acts as a carrier wave in radio.) However, it had never been done for light until 1960.

The laser was so designed, moreover, that the natural tendency of the photons to move in the same direction was accentuated. The two ends of the ruby tube were accurately machined and silvered so as to serve as plane mirrors. The emitted photons flashed back and forth along the rod, knocking out more photons at each pass, until they had built up sufficient intensity to burst through the end which was more lightly silvered. Those that did come through were precisely those that had happened to be emitted in a direction exactly parallel to the long axis of the rod, for only those would move back and forth, striking the mirrored ends over and over. If any photon of proper size happened to enter the rod in a different direction (even a very slightly different direction) and started a train of stimulated photons in that different direction, they would quickly pass out the sides of the rod after only a few reflections at most.

A beam of laser light is made up of coherent waves so firmly parallel that it can travel through long distances without widening to uselessness. Laser beams even reached to the moon, in 1962, having spread out to a diameter of only two miles after having crossed nearly a quarter of a million miles of space.

In the short time since their invention, lasers have proliferated in variety. They can be formed not only out of metallic oxides, but out of fluorides and tungstates, out of semiconductors, and out of columns of gas. Light can be produced in any of a variety of wavelengths in the visible and infrared ranges.

The narrowness of the beam of laser light means that a great deal of energy can be focused into an exceedingly small area, and in that area the temperature reaches extreme levels. The laser can vaporize metal for quick spectral investigation and analysis, and it can punch holes of any desired shape through high-melting substances. By shining laser beams into the eye, surgeons have succeeded in welding loosened retinas so rapidly that surrounding tissues have no time to be affected by the heat.

The possible applications of laser beams are exciting and dramatic, and they will probably come quickly. Rather than speculate about them now, it would be more appropriate to wait for later editions of this book and discuss them in actuality.

Matter-Waves

Bohr's application of quantum theory to atoms had thus proven incalculably fruitful, both in theory and application. Not only was the periodic table rationalized but a whole realm of solid-state devices had grown out of it. Physicists had every reason to be delighted with the results.

And yet, taken by itself, the quantized atom did not solve the problems of the chemist. It left him, for a while, with no clear-cut method for explaining the manner in which atoms clung together to form molecules. Where the Lewis-Langmuir atom, with all its faults and deficiencies, had enabled him to depict interlocking cubes and shared-electron pools, the quantized atom, with its electrons hopping nimbly from energy level to energy level, seemed impossible to handle.

An answer grew out of a second seeming source of confusion, that between particle and wave. The physicists of the early twentieth century had become convinced that light, and electromagnetic radiation generally, while wave form, also displayed particle-like properties. The Compton effect (see page II–138) had been the final convincer that wave-particle duality existed and that an entity could demonstrate both wave-like properties and particle-like properties.

But was this confined to electromagnetic radiation only? If entities commonly viewed as wave forms exhibited particle-like properties that can be detected if properly searched for, what of entities commonly viewed as particles? Would they exhibit wave-like properties that would be detected if properly searched for?

The French physicist Louis Victor de Broglie (1892–   ) considered this last problem. He made use of some of the relationships developed in treating a photon as a particle and applied them to electrons. In 1923, he announced the relationship:

$$\lambda = \frac{h}{mv}$$
    (Equation 6–1)

where h is Planck's constant (see page II–131), m is the mass of a moving particle, and v its velocity (and the product mv is its momentum). As for λ (the Greek letter "lambda"), that is the wavelength associated with its wave-like properties.

This equation will, in theory, apply to any moving body— to a baseball, a cannonball, a planet. However, as momentum increases, wavelength decreases, and for all ordinary bodies, the associated wavelength is far too small to be detected by any known method. Ordinary bodies can therefore be viewed as particles, without any worry about associated wave properties.

When the mass of an object decreases to that of an electron, however, the associated wavelength is significantly large—as large as that of an X ray. (The wave form associated with an electron is not, however, identical with an X ray in nature, though the wavelength may be the same. The wave forms associated with particles of matter are not electromagnetic in nature; these non-electromagnetic waves may be called *matter-waves*.)

A matter-wave with a wavelength equal to that of an X ray ought to have its wave nature as easily detectable as that of X rays. The wave nature of X rays was detected by the diffraction of X rays by the atom lattices of crystals (see page 63). Might not then the matter-waves associated with electrons be demonstrated in equivalent fashion?

This feat was independently carried out in 1927 by the American physicists Clinton Joseph Davisson (1881–1958) and Lester Halbert Germer (1896–   ) on the one hand, and by the English physicist George Paget Thomson (1892–   ) on the other. In later years, the wave properties of other, more massive, particles were also detected, and there is no reasonable doubt now that wave-particle duality is a general phenomenon in nature. All entities that display wave properties must also display particle properties, and vice versa.

The analogy between matter-waves and electromagnetic radiation showed up in the matter of microscopy.

There is a limit to the resolution possible when using a wave form such as light. Objects of a size less than about three-

fifths the wavelength of the light being used for the purpose cannot be made out, however perfect the optical portions of the
microscope. The light "steps over" the small object, so to speak.
This means that even when viewing with the shortest wavelengths
of visible light, say 380 millimicrons, an object less than 200
millimicrons in diameter cannot be made out. Viruses, which are
smaller than that, cannot therefore be seen by visible light, however one attempts to magnify. Indeed, the greatest useful magnification of an optical microscope is about 2000 times.

Successful attempts have been made to use electromagnetic
radiation of wavelength smaller than that of visible light, but
greater success was achieved with matter-waves. Electrons, with
an associated wavelength about equal to that of X rays, can be
used for the purpose. Electrons can be focused sharply by magnetic fields, just as light waves can be focused by lenses. A
specimen subjected to focused electrons must be quite thin to
allow the electrons to pass through; it must also be encased in a
good vacuum, otherwise the electrons will be scattered by air.
This limits the nature of objects that can be studied by electron
microscopy, but not too drastically.

The electrons, having passed through the specimen, form an
image on a fluorescent screen or on a photographic plate. Those
portions more opaque to electrons absorb and scatter them more
efficiently, and for this reason, a meaningful light-dark pattern is
produced.

The first *electron microscope* was prepared in Germany in
1931, the German electrical engineer Ernst August Friedrich
Ruske (1906–   ) being prominent in its development. By
1934, electron microscopes were developed that surpassed optical
microscopes in magnifying power, and by 1939 these were being
produced commercially. The magnifications made possible by
modern electron microscopes are a hundred times those within
the range of possibility of the best optical microscopes.

Matter-waves entered the realm of atomic theory, too. The
Austrian physicist Erwin Schrödinger (1887–1961) tackled the
problem of interpreting the structure of atoms in terms of particle
waves, rather than of particles alone.

Schrödinger pictured the electron as a wave form circling
the nucleus. It seemed to him that the electron could then exist
only in orbits of such size that the wave form occupies it in a
whole number of wavelengths. When this happens, the wave form
repeats itself as it goes round, falling exactly on itself, so to speak.
The electron is then a stable *standing wave*.

Electron microscope

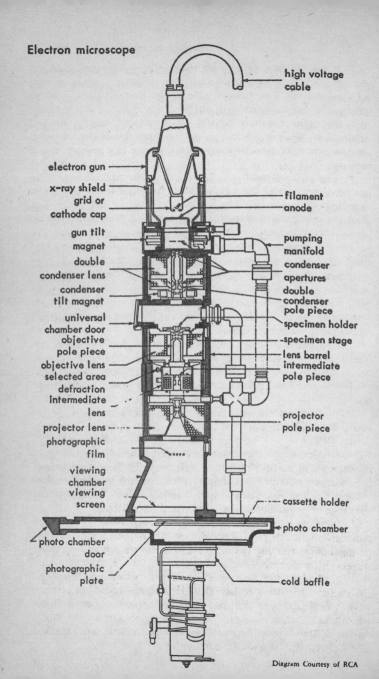

high voltage cable

electron gun

x-ray shield
grid or
cathode cap

filament
anode

gun tilt
magnet

pumping
manifold

double
condenser lens

condenser
apertures

condenser
tilt magnet

double
condenser
pole piece

universal
chamber door

specimen holder

objective
pole piece

specimen stage

objective lens

lens barrel
intermediate
pole piece

selected area
defraction
intermediate
lens

projector lens

projector
pole piece

photographic
film

viewing
chamber
viewing
screen

cassette holder

photo chamber

photo chamber
door

photographic
plate

cold baffle

Diagram Courtesy of RCA

If the electron gains a bit more energy, its wavelength decreases slightly and the orbit no longer contains a whole number of wavelengths. The same is true if the electron loses a bit of energy, so that its wavelength increases somewhat. If it is assumed that the electron cannot possess an amount of energy that will force it to circle a nucleus in a non-integral number of wavelengths, then the electron cannot gain or lose just any amount of energy.

The electron must gain (or lose) just enough energy to decrease (or increase) the wavelength to the point where an integral number of wavelengths can again fit the orbit. Instead of, say, four wavelengths to the orbit, there would be five somewhat shorter wavelengths to the orbit, with a gain of a specific quantity of energy, or three somewhat longer wavelengths, with a loss of a specific quantity of energy. If enough energy is lost and the wavelength increases to the point where a single wavelength fits the orbit, this is ground state and there can be no further loss of energy.

The different energy levels, then, represent different standing waves. Schrödinger analyzed this point of view mathematically in 1926, working out for the purpose what is now called the *Schrödinger wave equation*.

The analysis of the details of atomic behavior on the basis of the Schrödinger model is termed *wave mechanics*. Since the energy can only be absorbed or given off in quanta of given energy content, designed to maintain standing waves, it can also be called *quantum mechanics*.

Quantum mechanics has proved highly satisfactory from the physicist's standpoint. Psychologically, it seemed superior to Heisenberg's matrix mechanics (see page 90), for Schrödinger offered a picture, however hard to grasp, of wave forms, whereas Heisenberg's array of pictureless-numbers lacked something for the image-seeking mind to grasp.

In 1944, the Hungarian-American mathematician John von Neumann (1903–1957) presented a line of argument that seemed to show that quantum mechanics and matrix mechanics were mathematically equivalent—that everything that was demonstrated by one could be equally well demonstrated by the other.*

In principle, it would seem that quantum mechanics offers a complete analysis of the atom and that all facets of chemical be-

* In 1964, however, the English physicist Paul Adrien Maurice Dirac (1902–   ) raised doubts. He suggests the two theories are not mathematically equivalent and that matrix mechanics more accurately fits reality.

havior can be accounted for and predicted by means of it. In actual fact, however, a complete analysis is impractical, even by present-day techniques, because of the sheer difficulty of the mathematics involved. Chemistry is therefore far from being a completely solved science.

Nevertheless, quantum mechanics could be used to explain the manner in which atoms linked together to form molecules. The American chemist Linus Pauling (1901–   ) showed how two electrons could, in combination, form a more stable wave arrangement than they could separately. The shared-electron pool of the Lewis-Langmuir model of the atom became two wave forms resonating with each other (see page I–176). This *theory of resonance* was expounded fully in Pauling's book *The Nature of the Chemical Bond* published in 1939.

Resonance explains the structure and behavior of molecules far more satisfactorily than the old Lewis-Langmuir model does. It explains just those points that the older model left unexplained —such as the boron hydrides and benzene—and modern chemistry is more and more built about the quantum-mechanical viewpoint.

Another important consequence of the wave nature of the electron (and of particles generally) was pointed out by Heisenberg in 1927. You can see that if a particle is viewed as a wave, it is a rather fuzzier object than it would be if it were viewed as a particle only. Everything in the universe becomes slightly fuzzy, precisely because there is no such thing as a particle without wave-like properties.

A particle (or its center) can be located precisely in space —in principle, at least—but a wave form is somewhat harder to think of as being located at a particular point in space.

Thinking about this, Heisenberg advanced reasons for supposing that it is not possible to determine both the position and momentum of a particle simultaneously and with unlimited accuracy. He pointed out that if an effort is made to determine the position accurately (by any conceivable method, and not merely by those methods which are technically possible at the moment) one automatically alters the velocity of the particle, and therefore its momentum. Therefore, the value of the momentum at the moment at which the position was exactly determined becomes uncertain. Again, if one attempts to determine the momentum accurately, one automatically alters the position, the value of which becomes uncertain. The closer the pinning down of one, the greater the uncertainty in the other.

The concisest expression of this is:

$$(\Delta p)(\Delta x) \approx h \qquad \text{(Equation 6–2)}$$

where Δp represents the uncertainty of position, Δx the uncertainty of momentum, and h is Planck's constant. The symbol $\approx$ signifies "is approximately equal to." This is Heisenberg's *principle of uncertainty*.

Philosophically, this is an upsetting doctrine. Ever since the time of Newton, scientists and many nonscientists had felt that the methods of science, in principle at least, could make measurements that were precise without limits. One needed to take only enough time and trouble, and one could determine the nth decimal place. To be told that this was not so, but that there was a permanent wall in the way of total knowledge, a wall built by the inherent nature of the universe itself, was distressing.

Even Einstein found himself reluctant to accept the principle of uncertainty, for it meant that at the subatomic level, the law of cause and effect might not be strictly adhered to. Instead events might take place on the basis of some random effect. After all, an electron might be here or it might be there; if you couldn't tell, you couldn't be sure exactly how strongly a particular force at a particular point might affect it. "I can't believe," said Einstein, "that God would choose to play dice with the world."

Nevertheless, Einstein failed to devise any line of reasoning that would involve the principle of uncertainty in a contradiction. Nor could anyone else, and the principle is now firmly accepted by physicists.

Nor need one be overly downhearted about the loss of certainty. Planck's constant is very small, so for any object that is above the atomic in size, the relative uncertainties of position and momentum are vanishingly small. Only in the subatomic world need the principle be made a part of everyday life, so to speak.

What's more, the existence of uncertainty need not be a source of humiliation for science, either. If a tiny, but crucial, uncertainty is part of the fabric of the universe, it is a tribute to scientists to have discovered the fact. And surely, to know the limits of knowledge is itself an item of knowledge of the first importance.

CHAPTER 7

Radioactivity

Uranium

So far, the discussion of the internal structure of the atom has been confined to the outer electrons. In a way, it might almost seem that in doing so we were discussing virtually all the atom, for the nucleus has a diameter in the range of from 10^{-13} to 10^{-12} of a centimeter and makes up an insignificant portion of the atom. Indeed, if the atom were visualized as having been expanded to the size of the earth, the nucleus would be a sphere at the center of the planet, about 700 feet in diameter.

Yet the nucleus contains more than 99.9 percent of the mass of the atom, and almost from the first it was recognized (despite its minute size) as having an intricate structure of its own.

The first indication of this dates back to a discovery in 1896 by the French physicist Antoine Henri Becquerel (1852–1908). It was during the first year after Röntgen's discovery of X rays, and Becquerel, like many other physicists, was eagerly investigating the new phenomenon further.

Becquerel's father, himself a famous physicist, had been interested in fluorescent materials: substances which absorbed light of a particular wavelength and then gave off light of a longer wavelength.* Becquerel wondered if among the fluorescent radiation, there might not be X rays.

Becquerel's father had, in particular, worked with the fluorescent compound, potassium uranyl sulfate, $K_2UO_2(SO_4)_2$, the molecule of which, as you can see, contains a uranium atom. Becquerel, finding samples of this compound handy, used it in his experiments. He quickly discovered that after exposure to the sun the fluorescent radiation from the compound would penetrate black paper (opaque to ordinary light) and darken a photographic plate on the other side.

On March 1, 1896, however, he made the startling discovery that the compound would do this even when it had not been exposed to sunlight and when it was not fluorescing. Indeed, the compound constantly and ceaselessly emitted strong and penetrating radiation.

This radiation was not only as penetrating as X rays but, like X rays, possessed the ability to ionize the atmosphere. To demonstrate this, Becquerel made use of a *gold-leaf electroscope*. This device consists of two thin and very light sheets of gold leaf attached to a rod and enclosed in a box designed to protect the gold leaf from disturbing air currents. The rod emerges from the upper end of the box. If an electrically charged object is brought near the rod, the charge enters the gold leaf. Since both sheets of gold

---

* This is most spectacular when a fluorescent substance absorbs ultraviolet radiation and gives off light in the visible range. It seems, then, to glow eerily, and rather beautifully, in the dark.

Gold-leaf electroscope

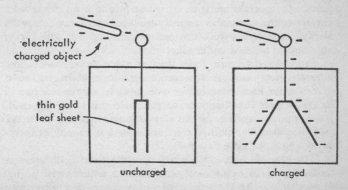

electrically charged object

thin gold leaf sheet

uncharged

charged

leaf are now similarly charged, they repel each other and stand stiffly apart, like an inverted V.

Left to itself, this situation will persist for an extended period of time. If, however, the air within the box is ionized, the charged particles in the air will gradually neutralize the charge on the gold leaf. The two sheets will then slowly come together as mutual repulsion vanishes. When potassium uranyl sulfate was brought near an electroscope, just this happened, so one could conclude that the compound liberated *ionizing radiation*.

This property of constantly emitting penetrating and ionizing radiation was termed *radioactivity* by the Polish-French physicist Marie Sklodowska Curie (1867–1934) in 1898. Madame Curie went on to show that different uranium compounds were all radioactive and that the intensity of radioactivity was in proportion to the uranium content of the compound. It seemed a fair conclusion that it was the uranium atom itself that was radioactive. Madame Curie was also able to show that the thorium atom was radioactive. (Both elements have particularly complex atoms. Thorium with an atomic number of 90 and uranium with one of 92 were the two most massive atoms known in the 1890's.)

It appeared almost at once that the radiation given off by uranium and by thorium was not homogeneous in its properties. In a magnetic field, part of the radiation was deflected in one direction by a very slight amount; part was deflected in the opposite direction by a considerable amount; and part remained undeflected. Ernest Rutherford (who was later to advance the nuclear model of the atom) gave these three parts of the radiation names taken from the first three letters of the Greek alphabet: *alpha rays*, *beta rays*, and *gamma rays*, respectively. These radiations also differed in respects other than their response to a magnetic field. Thus, it was the gamma rays which displayed the X ray-like penetrability. Beta rays were much less penetrating, and alpha rays were scarcely penetrating at all.

From the direction and extent of the beta ray deflection, Becquerel recognized that it must contain negatively-charged particles quite like those in cathode rays. He suggested this in 1899, and later investigations corroborated this over and over. Beta rays were shown to be streams of rapidly moving electrons. A speeding electron, emitted by a radioactive substance, is therefore commonly called a *beta particle*.

The gamma rays, undeflected by a magnetic field, were at once suspected of being electromagnetic in nature, with wavelengths even shorter than those of X rays since they were even more penetrating than X rays. This was unmistakably demonstrated in Rutherford's laboratory in 1914, when gamma rays were shown, like X rays, to be diffracted by crystals.

With the advent of the nuclear atom, it came to be realized that these radioactive radiations must originate out of events taking place within the nucleus. For instance, there are no energy level differences among the electrons of atoms which are large enough to produce photons as energetic as those of most gamma rays. Presumably there are *nuclear energy levels* within the nucleus, with differences large enough to produce gamma ray photons.

Yet the division between X rays and gamma rays is not a sharp one. While X rays, as a whole, have the longer wavelengths, some of the more massive atoms can produce X rays that are rather shorter in wavelength than some of the longest wave gamma rays originating from nuclei.

A wavelength of 0.01 millimicrons splits the overlap down the middle, so that as a rough rule of thumb it is possible to consider electromagnetic radiation on the short side of 0.01 millimicrons to be gamma rays and those on the long side to be X rays. The discovery of gamma rays completed the electromagnetic spectrum as we know it today. The stretch of radiation from the shortest gamma rays studied to the longest radio waves covers a range of some sixty octaves.

Alpha Particles

But what of the alpha rays? Their deflection in the direction opposite that of the beta rays showed that they must consist of positively-charged *alpha particles*. The fact that they were only slightly deflected by the same magnetic field that deflected the beta rays considerably made it quite likely that the alpha particles were much more massive than electrons.

This was not an unprecedented situation. Streams of massive particles were encountered a decade before the discovery of radioactivity. In 1886, Goldstein (who named the "cathode rays") had used a perforated cathode in a cathode-ray tube. He found that when an electric potential sent negatively-charged cathode rays streaking out of the cathode toward the anode, another type of radiation passed through the perforations of the cathode and shot off in the other direction. Goldstein called this second radiation "channel rays," because they passed through the channels, or holes, in the cathode.

Since the channel rays move in the direction opposite that of the cathode rays, they must consist of positively-charged particles. In consequence, J. J. Thomson suggested that they be called *positive rays*.

It might be suspected that the positive rays were the positive analog of the cathode ray particles; that here were the equivalents of "anode rays." Actually, this was not so. The German physicist Wilhelm Wien (1864–1928) measured their e/m ratio and showed that the low values of that ratio made it quite likely that the positive-ray particles were much more massive than electrons. They were, by and large, as massive as atoms.

Furthermore, the e/m ratio of the positive rays varied according to the nature of the substance making up the cathode, or according to the nature of the wisps of gas in the cathode-ray tube. Once Rutherford had developed his nuclear model of the atom, it seemed to make sense to suppose that where cathode rays consisted of electrons knocked out of atoms the positive rays consisted of what remained of atoms after some electrons had been removed. They were, in short, positively-charged atomic nuclei (varying in mass according to the element from which they were derived).*

The positively-charged particle that was found to have the highest e/m ratio and, therefore, presumably the lowest mass, was the nucleus of the hydrogen atom. If its charge is taken to be $+1$, equal to that of the electron but opposite in sign, then its mass had to be 1836 times as great as that of the electron. By 1914, Rutherford had given up hope of finding within the atom a positively-charged particle that was lighter than the hydrogen nucleus, and he suggested that this nucleus might as well be settled upon as the electron's opposite number, despite the difference in mass. (The discovery of the electron's true opposite number had to wait two more decades; see page 222.)

In 1920, Rutherford suggested that the hydrogen nucleus be given the name of *proton* (from a Greek word meaning "first").

---

* Of late, a dramatic use for such positively-charged particles has been forecast. The atom most easily stripped of at least some of its electrons is cesium, and, moreover, this atom is a comparatively massive one. A stream of cesium. ions, accelerated out of a rocket tube by an appropriate electric potential, would by Newton's third law (see page I-34) accelerate the rocket in the opposite direction. The force of a stream of even massive ions is small compared to the vast thrusts of the exhaust of burning fuel, but it can be long-continued. After chemical fuels do the heavy work of getting a rocket through earth's atmosphere and into outer space, an *ion-drive* may then prove the most economical method of slowly building up velocities near that of light. Perhaps only in this way can long space voyages become practical.

This harked back to Prout's hypothesis (see page 24), for what Rutherford was suggesting was that all atomic nuclei were made up, at least to a certain extent, of hydrogen nuclei. Prout's hypothesis was thus reborn in a more sophisticated form. The question of nonintegral atomic weights, which had seemed to destroy the hypothesis in the nineteenth century, was settled in a manner to be discussed later (see page 146).

Now let's return to the alpha particle. In 1906, Rutherford measured its e/m ratio and found that it was equivalent to that of the nucleus of the helium atom. In 1909, he settled this matter by placing radioactive material in a thin-walled tube, which was in turn surrounded by a thick-walled tube. The space between the inner and outer walls was evacuated. The alpha particles could penetrate the thin wall but not the thick one. After entering the space between the walls they picked up electrons and became ordinary atoms; then they could not pass through the thin wall either, but were trapped in the space between. After several days, enough atoms had been collected there to allow of spectroscopic investigation and the atoms proved to be those of helium.

The atomic weight of helium is 4, and the helium nucleus is therefore four times as massive as the hydrogen nucleus. If the e/m ratio of the helium nucleus were like that of the hydrogen nucleus, the helium nucleus would have to have a positive charge four times that of the hydrogen nucleus. However, the e/m ratio of the helium nucleus is only half that of the proton, so that its electric charge is only half the expected amount, or only twice that of the hydrogen nucleus. The alpha particle (as we can fairly term the helium nucleus) has, therefore, a mass of 4 and a charge of $+2$, whereas the proton (or hydrogen nucleus) has a mass of 1 and a charge of $+1$.

It would seem from this that to account for its mass the alpha particle must consist of four protons. Yet it cannot consist of four protons only, for then its charge would be $+4$. There seemed, however, an easy solution to this apparent paradox. Since radioactive substances emitted beta particles (electrons) as well as alpha particles, it seemed quite reasonable to suppose that the nucleus contained electrons as well as protons. The alpha particle, from this point of view, could be made up of four protons and two electrons. The two electrons would add virtually nothing to the mass, which would remain 4, but they would cancel the charge on two of the protons, leaving a net charge of $+2$.

The existence of electrons in the nucleus also seemed satisfactory from another standpoint. The nucleus could not very well consist of protons only, it seemed, for all the protons would be positively charged and there would be a colossally strong repulsion among them when forced into the ultra-narrow confines of an atomic nucleus. The presence of the negatively-charged electrons acted as a kind of "cement" between the protons.

Considerations of this sort gave rise to the *proton-electron model* of the atomic nucleus. Every nucleus, according to this view, was made up of both protons and electrons (except the hydrogen nucleus, which was made up of a single proton requiring, since it was alone, no electron-cement).

The number of protons in each variety of nucleus would be equal to the atomic weight (A),* while the number of electrons was equal to the number required to cancel the charge of enough protons to leave uncanceled only the amount required to account for the atomic number (Z). The number of electrons in a nucleus would, therefore, be equal to $A - Z$. Those protons remaining uncanceled in the nucleus would have their charge canceled by the electrons outside the nucleus, so that in the neutral atom there would be Z "extra-nuclear electrons."

Thus, to give a few examples, the nucleus of the carbon atom, which has an atomic weight of 12 and an atomic number of 6, must be made up of twelve protons and $12 - 6$, or six electrons. The nucleus of the arsenic atom with an atomic weight of 75 and an atomic number of 33, must be made up of seventy-five protons and $75 - 33$, or forty-two electrons. The nucleus of the uranium atom with an atomic weight of 238 and an atomic number of 92 must be made up of 238 protons and $238 - 92$, or 146 electrons. Even the nucleus of the hydrogen atom fits this view, for with an atomic weight of 1 and an atomic number of 1, it must be made up of one proton and $1 - 1$, or zero electrons.

Unfortunately, the proton-electron model of the atomic nucleus met with difficulties. For example, there is the question of *nuclear spin*. Each particle in the nucleus contributes its own spin to the overall nuclear spin. The spin of each proton and electron is either $+1/2$ or $-1/2$, and the sum of a number of such values can be either a whole number (positive, negative, or zero) or a half-number such as $1/2$, $3/2$, $5/2$, etc. (either positive or negative).

The nitrogen nucleus, with an atomic weight of 14 and an atomic number of 7, ought, by the proton-electron model, have

* Strictly speaking, this applies only to those elements whose atomic weights are approximately whole numbers. The situation with respect to the other elements will be considered later.

fourteen protons and seven electrons for a total of 21 particles in the nucleus. Therefore, if the spins of 21 particles (each $+1/2$ or $-1/2$) are totaled, regardless of the distribution of negative or positive spins, the sum must be a half-number. Nevertheless, measurements convinced physicists that the spin of the nitrogen nucleus was the equivalent of a whole number. This gave them good reason to suppose that the nitrogen nucleus could not be made up of twenty-one protons and electrons, or indeed of any odd number of protons and electrons. Yet no even number of protons and electrons could produce an atomic weight of 14 and an atomic number of 7.

It seemed more and more necessary, as additional data on nuclear spins were accumulated, for the proton-electron model to be scrapped altogether.

Particle Detection

Yet what alternative was there? One possibility was that the electron within the nucleus ought not be counted as a separate particle. Perhaps, in the close confines of the tiny nucleus, the electron melted together with a proton to form a single "fused particle," with a mass of about that of a proton (since the electron contributes very little mass) and with an electric charge of 0 (since the electron's charge of -1 cancels the proton's charge of $+1$). If this were so then a nitrogen nucleus would contain seven protons plus seven "fused particles" for 14 particles altogether—an even number.

Speculations concerning the possible existence of uncharged particles with the mass of a proton began as early as 1920. For over a decade, however, no signs of such a particle could be found. This did not necessarily mean that it did not exist, for physicists expected an uncharged particle to be elusive.

The usual methods for detecting a subatomic particle took advantage of its ability to ionize atoms and molecules. It was by their ionizing abilities, for instance, that radioactive radiations were detected by the electroscope.

Two devices, in particular, used for the purpose of detecting subatomic particles, grew famous in the early days of research into radioactivity. The prototype of the first of these was constructed in 1913 by the German physicist Hans Geiger (1882–1945), who had worked with Rutherford in the experiments that led to the working out of the nuclear model of the atom. It was greatly improved by Geiger in 1928 in collaboration with the German physicist S. Müller, and is therefore commonly known as the *Geiger-Müller counter*, or the *G-M counter*.

The G-M counter consists, essentially, of a cylindrical glass tube lined with metal, with a thin metal wire running down the center of the tube. Filling the tube is a gas such as argon. The tube is placed under an electric potential, with the central wire the anode and the metal cylinder as cathode. The potential is not quite great enough to cause a spark discharge across the argon.

If a charged subatomic particle comes flying into the G-M counter, it will strike an argon atom and knock one or more electrons loose. The electrons so formed will then speed toward the anode under the lash of the electric potential and will in turn ionize other argon atoms, producing more electrons which will in turn ionize still other argon atoms, and so on. In short, the first particle starts a process that in a small instant of time produces so many ions that the argon becomes capable of conducting a current, and there is then an electric discharge within the tube that momentarily reduces its potential to zero.

The discharge, or pulse, of electric current can be converted into a clicking sound that marks the passage of a single subatomic particle. From the number of clicks per second one can estimate, by ear, how much ionizing radiation is present. (Hence G-M counters are used in uranium prospecting.) The pulses can also be counted accurately by automatic devices.

To do more than merely count subatomic particles, one can use a device invented in 1911 by the Scottish physicist Charles Thomas Rees Wilson (1869–1959). He was primarily interested in cloud formation to begin with and had come to the conclusion that water droplets in clouds were formed about dust particles and could also form about ions. If air were completely free of dust or ions, clouds would not form and the air would become *supersaturated*—that is, it would retain water vapor in quantities greater than it could ordinarily hold.

Wilson placed dust-free air, saturated with water vapor, into a chamber fitted with a piston. If the piston were pulled outward, the air would expand and its temperature would drop. Cold air cannot hold as much water vapor as warm air can, and ordinarily some of the water vapor would have to condense out in drops of liquid as the temperature dropped. In the absence of dust or ions, this could not happen and the cooled air became supersaturated.

If, now, a subatomic particle entered the chamber while the air was supersaturated, it would form ions along its path of travel, and small droplets of water would form about these ions. These droplets would mark out the route taken by the subatomic particle.

The tracks formed in such a *Wilson cloud chamber* are packed with information. Different types of particles can be identified. A massive alpha particle, for instance, forms many ions in its path and continues in a straight line, for it is too massive to be diverted by electrons. It is diverted only when it approaches a nucleus, and then the diversion is likely to be a sharp one. The nucleus, stripped of some electrons, recoils and becomes an ionizing particle itself. The track of an alpha particle is therefore thick and straight, and it usually ends with a fork. From the length of the track one can estimate the original energy of the alpha particle.

A beta particle, which is much lighter, changes its direction of motion more easily and forms fewer ions than an alpha particle does. It leaves a thinner and more wavering track. Gamma rays and X rays knock electrons out of atoms, and these act as ionizing particles that mark out short tracks to either side of the path of the gamma ray or X ray. Such radiation therefore leaves faint fuzzy tracks.

If a cloud chamber is placed between the poles of a magnet, charged particles travel in curved paths, and the water droplets indicate that. From the direction of curve, one can determine whether the charge is negative or positive; and from the sharpness of the curve, one can make deductions as to the e/m ratio.

For a speeding particle to form ions, however, the presence of an electric charge is essential. A positively-charged particle attracts electrons out of the atoms it passes through, and a negatively-charged particle repels them out of the atom. An uncharged

Wilson cloud chamber

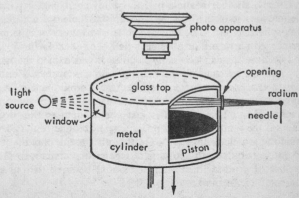

particle would neither attract nor repel electrons and would not form ions. Such an uncharged particle would, therefore, go unmarked by either a G-M counter or a Wilson cloud chamber (or, for that matter, by the more sophisticated devices that have been invented since). If an uncharged particle exists at all, therefore, it would have to be detected indirectly.

It was this which stood in the way of the easy detection of a neutral particle and delayed for over a decade the development of a nuclear model more satisfactory than the proton-electron model.

The Neutron

Beginning in 1930, evidence was obtained to the effect that when beryllium was exposed to alpha rays a radiation was emitted that differed from those that were already known. It was very penetrating, and it was not affected by a magnetic field, which seemed to give it the hallmark of gamma rays. However, the radiation was not gamma ray in nature either, for unlike gamma rays, it was not an ionizing radiation and could not, for instance, be detected by an electroscope.

Indeed, the radiation was not detected directly at all. When a substance such as paraffin was put in the way of the radiation, however, protons were hurled out of the paraffin and these protons gave it away.

In 1932, the English physicist James Chadwick (1891–   ) explained the phenomenon satisfactorily. Energetic electromagnetic radiation might push electrons out of the way, but not the more massive protons. For a proton to be banged about so cavalierly, another particle had to be involved, and a particle in the mass-range of the proton. Since this particle did not ionize the atmosphere, it had to be uncharged. In short, here was the massive uncharged particle that physicists had been seeking for a decade, and because it was electrically neutral, it was named the *neutron*.

Almost as soon as the neutron was discovered, Heisenberg suggested a *proton-neutron model* of the atomic nucleus. According to this model, the nucleus would be made up of protons and neutrons only. The neutron is just about equal to the proton in mass, so the total number of protons (p) and neutrons (n) would equal the atomic weight (A). On the other hand, only the protons would contribute to the positive charge of the nucleus, so the total number of protons in the nucleus would be equal to the atomic number (Z). In short:

$$p + n = A \qquad \text{(Equation 7–1)}$$

$$p = Z \qquad \text{(Equation 7–2)}$$

The number of neutrons, then, can be obtained by subtracting Equation 7–2 from Equation 7–1·

$$n = A - Z \qquad \text{(Equation 7–3)}$$

Using this new view, it is easy to specify the structure of the nuclei of those atoms that have atomic weights that are approximately whole numbers.

The nucleus of the hydrogen atom $(A = 1, Z = 1)$ is made up of one proton and zero neutrons; that of the helium atom $(A = 4, Z = 2)$ is two protons and two neutrons; that of the arsenic atom $(A = 75, Z = 33)$ is thirty-three protons and forty-two neutrons; and that of the uranium atom $(A = 238, Z = 92)$ is ninety-two protons and one hundred and forty-six neutrons.

This proton-neutron model of the nucleus quickly showed itself satisfactory in those respects in which the proton-electron model failed. The nitrogen nucleus, for instance $(A = 14, Z = 7)$ is made up of seven protons and seven neutrons for a total of 14 particles, altogether. The neutron, as well as the proton, has a spin of $+1/2$ or $-1/2$, and 14 such particles (or any even number) must have a net spin represented by a whole number, in agreement with observation.

The proton-neutron model is still accepted at the present writing, and the two particles are lumped together as *nucleons* because of their characteristic appearance in atomic nuclei.

The model does, of course, raise questions. One of these is this: If the nucleus contains protons and neutrons only, and does not contain electrons, then where do the electrons come from that make up the beta rays emitted by radioactive substances? It was, after all, the existence and nature of these beta rays that led to the belief in nuclear electrons in the first place.

The answer to this arises out of the nature of the neutron which, in one particular respect, differs crucially from the proton and the electron. Both the proton and the electron are examples of a *stable particle*. That is, if either a proton or an electron were alone in the universe it would persist, unchanged, indefinitely (at least as far as we know). Not so with the neutron, which is an *unstable particle*.

An isolated neutron will eventually cease to exist, and in its place will be two particles, a proton and an electron. (This is not a complete description of the breakdown, but it will do for now.

The subject will be explored further on page 236.) We can write this change symbolically, using superscripts to indicate charge, as follows:

$$n^o \longrightarrow p^+ + e^-$$ (Equation 7–4)

An important point demonstrated by this equation is that electric charge is not created. All experience involving the behavior of subatomic particles indicates that the neutron cannot merely change into a proton, for there would be no way in which an uncharged particle could develop a positive charge (or, for that matter, a negative charge) out of nothing. By forming both a proton and an electron, the net charge of the product remains zero.

This is an example of the *law of conservation of electric charge*, which states that the net charge of a closed system cannot be altered by changes taking place within the system. This was first recognized in the study of electrical phenomena (see page II–161), long before the existence of subatomic particles was suspected.

A neutron existing, not free, but within an atomic nucleus, is often stabilized for reasons to be discussed later (see page 243). Thus, the nucleus of a nitrogen atom is stable despite its neutron content and, left to itself, will continue to be made up of seven protons and seven neutrons indefinitely.*

On the other hand, there are some nuclei within which neutrons retain a certain degree of instability. In such cases, the neutron within the nucleus will, at some point, change into a proton and an electron. The proton is at home in the nucleus and remains there, but the electron comes flying out as a beta particle. Thus, although the beta particle does emerge from the nucleus, this is no indication, after all, that it was a constituent of the nucleus; rather, it was created at the moment of its emergence.

New Radioactive Elements

If, in the course of the emission of a beta particle, a neutron within the nucleus of an atom is converted into a proton, it is clear that the proton-neutron makeup of the nucleus changes and that the nature of the atom itself is altered. Since the number of

* It might occur to you to wonder what holds the seven positively-charged protons together against the powerful electric repulsion of like charges, in the absence of an "electron-cement." This problem will be considered later in the book (see page 243).

protons is increased by one, so is the atomic number, and the atom within which the change has taken place is transformed from one element into another.

In fact, radioactivity is almost invariably a sign of a fundamental change in the nature of the atom displaying the phenomenon. This came to be realized shortly after the discovery of radioactivity, and well before the internal structure of the nucleus was worked out.

As early as 1900, Crookes, one of the cathode-ray pioneers, discovered that when a uranium compound was thoroughly purified it showed virtually no radioactivity. It was his suggestion, therefore, that it was not uranium that was radioactive, but some impurity in the uranium.

However, the next year Becquerel confirmed Crookes' findings but went on to show that as the purified uranium compound remained standing, its radioactivity gradually grew more intense, until it was at the normal level associated with uranium. In 1902, Rutherford and his co-worker, the English physicist Frederick Soddy (1877–1956), showed this to be also true of thorium compounds.

It seemed reasonable to conclude that if the radioactivity was that of an impurity, it was an impurity that was gradually formed from uranium. In other words, the radioactivity of uranium was a symptom of the change of uranium atoms into some other form of atom. This new atom was itself radioactive and changed into a third atom which was also radioactive, and so on. In short, as Rutherford and Soddy pointed out, one ought not speak of a radioactive element, but of a *radioactive series* of elements.

The radioactivity detected in uranium and thorium might not then be so much characteristic of uranium and thorium itself (which might be, and indeed proved to be, only very mildly radioactive) as of the various "daughter elements." The latter were much more strongly radioactive and were always present in the uranium and thorium—except immediately after those elements had been rigorously purified.

The "daughter elements," if slowly formed and rapidly broken down, ought to be present in uranium and thorium minerals only in vanishingly small quantities. Even so, while they would remain immune to discovery by ordinary chemical methods, they could be detected and traced by the radiations they gave off, since these could be detected with great sensitivity and since it was only to be expected that each different element would give off radiations of characteristic type and intensity.

This feat was successfully carried through by Madame Curie, in collaboration with her husband, the French physicist Pierre Curie (1859–1906). In 1898, the Curies began with large quantities of uranium ore and divided it, by standard chemical techniques, into fractions of different properties. They followed the track of intense radioactivity, keeping those fractions that displayed it and discarding those that did not. Before the end of the year, they had discovered two hitherto unknown elements, the first of which they named *polonium* after Madame Curie's native land, and the second, *radium*, after the element's intense radioactivity.

Both elements were, in fact, far more radioactive than either uranium or thorium. In fact the rapidity with which polonium and radium broke down and ceased being polonium and radium was such that no detectable quantity could possibly have survived the five-billion-year history of the earth, even if large quantities had existed in the planet's structure when it was formed. The existence of these elements today was due entirely to their constant formation from uranium and thorium. The latter elements broke down so slowly that a sizable fraction of the original supply remains in existence today, despite a steady diminution over the last five billion years.

How many such short-lived elements might exist as daughter products of uranium and thorium? In the time of the Curies this was uncertain, for there was no telling how much room there might remain in the periodic table. Once Moseley had worked out the concept of atomic numbers, in 1913, the subject grew less mysterious.

As of 1913, all elements with atomic numbers up to and including 83 (bismuth) were nonradioactive. It was thoroughly expected that the as-yet-undiscovered elements in this range (43, 61, 72, and 75) would also be nonradioactive. And, to be sure, when hafnium (72) was discovered in 1923 and rhenium (75) in 1925, both turned out to be nonradioactive. Attention, then, was focused on elements of atomic number higher than 83.

Thorium (atomic number 90) and uranium (atomic number 92) were the first radioactive elements discovered. Those discovered by the Curies also fit into this region, for polonium had an atomic number of 84 and radium one of 88.

Other discoveries followed. In 1899, the French chemist André Louis Debierne (1874–1949) discovered *actinium* (atomic number 89), and in 1900, the German chemist Friedrich Ernst Dorn (1848–1916) discovered *radon* (atomic number 86). In

1917, the German chemist Otto Hahn (1879–   ) and his co-worker, the Austrian physicist Lise Meitner (1878–   ), discovered protactinium (atomic number 91).

By that time (and for a quarter-century afterward) only two gaps remained in that region of the periodic table, gaps corresponding to atomic numbers 85 and 87. Chemists confidently expected the elements with those atomic numbers to prove radioactive when discovered (and this turned out to be so).

And yet, as we shall see in the next chapter, this listing of elements, which seems to fit the periodic table so neatly, actually involved chemists in a problem that began by seeming to shake the very concept of a periodic table and ended by establishing it more firmly and more fruitfully than ever.

CHAPTER 8

Isotopes

Atomic Transformations

The discovery of new elements in radioactive minerals in 1898 and immediately thereafter was, in a way, too successful for comfort. The periodic table had room for exactly nine radioactive elements with atomic numbers of from 84 to 92 inclusive. Room for new elements such as radium and polonium could be found, but how many more were there? If one judged by the number of distinct and characteristic types and the intensities of radiation among the daughter elements of uranium and thorium, then physicists seemed to have discovered dozens of different elements.

Names were applied to each distinct type of radiation: There were, for instance, uranium X_1, uranium X_2, radium A, radium B, and so on through radium G. There was also a list of thoriums from A to D, two mesothoriums, a radiothorium, and so on. But if each type of radiation did indeed belong to a different element, where could one place them all? Once Moseley had worked out the atomic number structure of the periodic table, the problem had grown crucial.

To answer this problem, let's consider the nature of the radioactive radiations and the manner in which they must affect the atom giving them off. (I will make use of the proton-neutron model of the atomic nucleus, though the analysis I will describe was worked out originally on the basis of the proton-electron model.)

Let's begin with an element, Q, and suppose that its nucleus is made up of x protons and y neutrons. Its atomic number, then, is x and its atomic weight is $x + y$. Placing the atomic number as a subscript before the symbol of the element and the atomic weight as a superscript after it, we can write the element as $_xQ^{x+y}$.

Next let us suppose that an atom of this element gives off an alpha particle (symbolized by the Greek letter α, which is "alpha"). The alpha particle is made up of two protons and two neutrons and therefore has an atomic number of 2 and an atomic weight of 4. It can be written: $_2\alpha^4$.

What is left of the original atom after the departure of an alpha particle must contain $x - 2$ protons and $y - 2$ neutrons. The atomic number is decreased by 2 (producing a new element, R), and the atomic weight is decreased by 4. We can write this:

$$_xQ^{x+y} \longrightarrow {}_{x-2}R^{x+y-4} + {}_2\alpha^4 \qquad \text{(Equation 8–1)}$$

If the original atom had given off a beta particle (symbolized as β, the Greek letter "beta") instead, the situation would be different. The emission of a beta particle means that within the nucleus a neutron had been converted into a proton. The nucleus would therefore contain $x + 1$ protons and $y - 1$ neutrons. The atomic number would be increased by one, but the atomic weight would remain unchanged for $x + 1 + y - 1 = x + y$.

The beta particle itself can be considered as having an atomic weight of about 0. (Actually, it is 0.00054, which is close enough to zero for our purposes.) Since atomic number is equivalent to the number of units of positive charges present and since the beta particle is an electron with a unit negative charge, we can consider its atomic number as equal to -1. The beta particle can therefore be written as $_{-1}\beta^0$ and beta particle emission can be represented as:

$$_xQ^{x+y} \longrightarrow {}_{x+1}R^{x+y} + {}_1\beta^0 \qquad \text{(Equation 8–2)}$$

Notice that in both Equation 8–1 and Equation 8–2, the sum of the atomic numbers on the right-hand side of the equation is equal to that on the left-hand side. This is in accordance with the law of conservation of electric charge. The same is true for atomic weights in accordance with the law of conservation of mass. (The minor deviations involving the conversion of some mass into energy need not concern us just yet.)

A gamma ray can be symbolized as γ, the Greek letter

"gamma." Since it is electromagnetic radiation, it has neither atomic weight nor atomic number and can be written $_0\gamma^0$. We can, therefore, write the following equation:

$$_zQ^{x+y} \longrightarrow _zQ^{x+y} + _0\gamma^0 \qquad \text{(Equation 8–3)}$$

In short, then, when an atom emits an alpha particle, its atomic number decreases by two and its atomic weight decreases by four. When it emits a beta particle, its atomic number increases by one and its atomic weight is unchanged. When it emits a gamma ray, its atomic number and atomic weight are both unchanged. This is the *group displacement law*, first proposed in its complete form by Soddy in 1913.

Let us apply the group displacement law to the specific case of the uranium atom with an atomic number of 92 and an atomic weight of 238—that is, $_{92}U^{238}$. The feeble radioactivity of highly purified uranium consists of alpha particles. An alpha particle emission reduces the atomic number of uranium to 90, which is that of thorium, and reduces its atomic weight to 234. We can write:

$$_{92}U^{238} \longrightarrow _{90}Th^{234} + _2\alpha^4 \qquad \text{(Equation 8–4)}$$

The thorium atom that has arisen as a result of this breakdown of the uranium atom is not quite like the thorium atom that occurs in quantity in thorium ores. The latter possesses an atomic number of 90, to be sure, but it has an atomic weight of 232. It is $_{90}Th^{232}$.

Both types of thorium atoms possess the atomic number of 90 and fit into the same place in the periodic table. Soddy pointed this out in 1913 and suggested that atoms differing in atomic weight but not in atomic number, as in the case of $_{90}Th^{234}$ and $_{90}Th^{232}$, be referred to as *isotopes*, from Greek words meaning "same place," because they occupy the same place in the periodic table.

Since such isotopes always share the same atomic number and differ in atomic weight only, chemists concentrate on the latter and usually leave out the subscript in writing the symbol of the isotope. They will write the two thorium isotopes as Th^{234} and Th^{232} or, less compactly, as thorium-234 and thorium-232.

From the chemist's point of view the placing of different isotopes in the same place in the periodic table is justified. Both thorium-234 and thorium-232 have 90 protons in the nucleus and therefore 90 electrons outside the nucleus in the neutral atom. The chemical properties depend on the number and distribution of the electrons, and therefore thorium-234 and thorium-232 have virtually identical chemical properties. This reasoning holds for other sets of isotopes as well.*

But if different isotopes have identical complements of outer electrons, they nevertheless have nuclei of different structures. Since the proton number in the nuclei of different isotopes of an element is fixed, the difference must rest in the neutron number. The thorium-234 atom, for instance, has a nucleus made up of 90 protons and 144 neutrons, while the thorium-232 atom has one made up of 90 protons and 142 neutrons.

In the case of changes involving the atomic nucleus, such as those that mark the phenomenon of radioactivity (as contrasted with chemical changes that involve only the electrons and not the nucleus), the difference in neutron number is important.

Thus, thorium-232 breaks down with exceeding slowness, which is exactly why it is still present in the crust. It emits an alpha particle, so that its atomic number is reduced to 88, which is that of radium. We can write:

* There are some minor differences in chemical property, particularly among the lighter atoms, because one isotope is more massive than another and therefore somewhat more sluggish in taking part in chemical reactions. These differences are small, however, and in ordinary chemical work can be ignored.

Simple isotopes

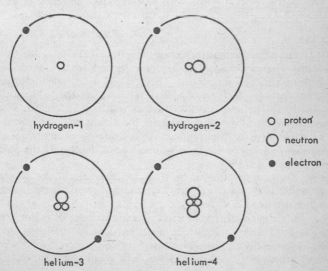

hydrogen-1	hydrogen-2
helium-3	helium-4

○ proton
◎ neutron
● electron

$$_{90}\text{Th}^{232} \longrightarrow {}_{88}\text{Ra}^{228} + {}_{2}\alpha^{4} \qquad \text{(Equation 8-5)}$$

Atoms of thorium-234, on the other hand, break down with exceeding rapidity, which is why this isotope does not occur in nature except in vanishingly small quantities in uranium ores. Furthermore, it breaks down with the emission of a beta particle. This raises its atomic number to 91, that of protactinium:

$$_{90}\text{Th}^{234} \longrightarrow {}_{91}\text{Pa}^{234} + {}_{-1}\beta^{0} \qquad \text{(Equation 8-6)}$$

Where either an alpha particle or a beta particle is emitted, the new atom may be formed at a nuclear energy level above the ground state. In dropping to the ground state thereafter, a gamma ray is emitted. In some cases this takes an appreciable time and the excited nucleus has a lifetime of its own and different radiation characteristics. To indicate a nucleus in the excited state, an asterisk is added to the symbol. When protactinium-234 is formed, it is in the excited state and:

$$_{91}\text{Pa}^{234*} \longrightarrow {}_{91}\text{Pa}^{324} + {}_{0}\gamma^{0} \qquad \text{(Equation 8-7)}$$

Atoms that are identical in atomic weight and atomic number but differ in nuclear energy level are called *nuclear isomers,* a name suggested by Lise Meitner in 1936. The first evidence for nuclear isomerism had been obtained by her long-time partner, Otto Hahn, in connection with protactinium-234, back in 1921.

Radioactive Series

Once the group displacement law was worked out, the trivial names given to the different atoms formed from uranium and thorium could be abandoned. They are still to be found in physics books because of their historical interest, but they will not be used here. The proper isotope names will be used instead. When this is done it turns out that despite the dozens of isotopes formed in the course of the radioactive breakdown of uranium and thorium, all can be made to fit into one or another of the places in the atomic table.

You can see this to be true, for instance, of the different atoms formed from uranium-238, the so-called *uranium series,* which are listed in Table V.

If we consider this series in detail, a number of points arise. First, lead-206 is a stable isotope that does not undergo radioactive breakdown. The series therefore ends there. Nevertheless, there are also included in the series such lead isotopes as lead-214 and lead-210, which are radioactive. Here is a clear indication that isotopes are a phenomenon that are not confined to radioactive atoms alone, but that a particular element may have both stable and radioactive isotopes.

If we leave lead-206 out of consideration and take into account just the radioactive members of the series, only uranium-238 breaks down with exceeding slowness. All the rest break down with comparative rapidity. Consequently, only uranium-238 can endure over the full stretch of the earth's existence. It is the "parent" of the series, and none of the daughter atoms would exist on earth today if uranium-238 did not.

TABLE V— *The Uranium Series*

uranium-238
$\downarrow -\alpha$
thorium-234
$\downarrow -\beta$
protactinium-234
$\downarrow -\beta$
uranium-234
$\downarrow -\alpha$
thorium-230
$\downarrow -\alpha$
radium-226
$\downarrow -\alpha$
radon-222
$\downarrow -\alpha$
polonium-218

$-\alpha \swarrow \qquad \searrow -\beta$

lead-214 $\qquad\qquad$ astatine-218

$-\beta \searrow \quad$ bismuth-214 $\quad \swarrow -\alpha$

$-\beta \swarrow \qquad\qquad \searrow -\alpha$

polonium-214 $\qquad\qquad$ thallium-210

$-\alpha \searrow \quad$ lead-210 $\quad \swarrow -\beta$

$\downarrow -\beta$

$-\beta \swarrow \quad$ bismuth-210 $\quad \searrow -\alpha$

polonium-210 $\qquad\qquad$ thallium-206

$-\alpha \searrow \quad$ lead-206 $\quad \swarrow$

A particular radioactive atom need not always have only one mode of breakdown. Polonium-218, for instance, may give off an alpha particle to form lead-214, or it may give off a beta particle to form astatine-218. This is an example of *branched disintegration*. Often, in these cases, one of the branches is the overwhelming favorite. For instance out of every 10,000 atoms of polonium-218, only two break down to astatine-218, all the rest breaking down to lead-214. (In this case, it is the alpha particle emission that is the favored alternative, in other cases the beta particle emission is favored.)

Astatine (atomic number 85), when formed at all in radio-active breakdowns, is usually formed at the very short end of a branched disintegration. That is why it exists naturally in almost unimaginably small traces and why it evaded discovery so long. The same is true of francium (atomic number 87), which is not formed at all in this particular series.

In a radioactive series, the atomic weight of any atom faces one of two fates. Either its value does not change at all, when a beta particle or a gamma ray is emitted, or it decreases by 4 units.

TABLE VI—*The Thorium Series*

thorium-232
↓ −α
radium-228
↓ −β
actinium-228
↓ −β
thorium-228
↓ −α
radium-224
↓ −α
radon-220
↓ −α
      −α polonium-216 −β

lead-212   astatine-216
      −β   −α
         bismuth-212
      −α   −β
thallium-208   polonium-212
      −β   lead-208   −α

when an alpha particle is emitted. That means that the difference in atomic weights between any two members of the series must be 0 or else a multiple of 4.

The atomic weight of uranium-238 is 238 and this, when divided by 4, gives a quotient of 59 with a remainder of 2. For any number that differs from 238 by a multiple of 4, division by 4 will yield a different quotient but will always leave a remainder of 2. The value of the atomic weight of every member of the uranium series therefore has the form of $4x + 2$, where x can vary from 59 for uranium-238 down to 51 for lead-206. For this reason, one can call the uranium series the *$4x + 2$ series*.

Thorium, the second element to be discovered to be radio-active, is also the parent of a group of daughter atoms, the *thorium series* (see Table VI).

Here, too, the atomic weight of, all the atoms in the series differs by multiples of 4. Since thorium-232 has an atomic weight, 232, which is evenly divisible by 4, all the other atomic weights in the series must be evenly divisible by 4, and the series may be referred to as the *$4x + 0$ series*.

It might be thought that since uranium and thorium are the only two radioactive elements that occur in appreciable quantities in the soil, there would be only two radioactive series. However, atoms appeared in radioactive minerals with atomic weights that were neither of the form $4x + 0$ nor $4x + 2$ and which therefore could belong to neither the thorium series nor the uranium series.

At first, it was felt that these formed part of a series originating with actinium-227, an isotope with a $4x + 3$ type of atomic weight, so it was named the *actinium series*. The name persists even though this supposition was sent tumbling by the discovery that actinium-227 broke down far too quickly to permit its existence through the eons of earth's history, so that it could not possibly serve as the parent atom of a series.

When protactinium was discovered, it was found that protactinium-231 (to use present terminology) broke down to form actinium-227, and that, indeed, was the reason the new element received its name (which means "before actinium"). Protactinium-231, however, is also too short-lived to qualify as parent of a series.

In 1935, the Canadian-American physicist Arthur Jeffrey Dempster (1886–1950) discovered that not all uranium atoms were uranium-238. Out of every thousand atoms of uranium isolated from the natural ore, seven were *uranium-235*. These atoms, possessing 92 protons and 143 neutrons in their nuclei, broke down very slowly (though not quite as slowly as uranium-238) and qualified as the parent atom of the actinium series. (For this reason, uranium-235 is sometimes called "actinouranium.") The atomic weights of all the atoms making up the actinium series (see Table VII) are of the $4x + 3$ variety.

It is plain that still a fourth radioactive series should exist, one in which all the atomic weights are of the form $4x + 1$. Isotopes of this form, such as uranium-233, cannot belong to any of the three series already described. No such series was discovered in the 1920's and 1930's, and physicists decided (correctly, as it turned out) that no isotope with an atomic weight of this form was long-lived enough to serve as a parent for such a series in nature.

Each of the three radioactive series ends with a lead isotope with an atomic weight of appropriate form. The uranium series ends with lead-206 ($4x + 2$), the thorium series ends with lead-208 ($4x + 0$), and the actinium series ends with lead-207 ($4x +$

TABLE VII—The Actinium Series

3). All three lead isotopes are stable, which indicates that an element may not only contain both stable and unstable isotopes but more than one stable isotope as well.

Half-Lives

So far I have talked about radioactive isotopes that underwent radioactive breakdown very slowly and others that broke down rapidly, but I have made no attempt to assign actual figures to these qualitative descriptions.

The first attempt to do so was made by Rutherford and Soddy in investigations beginning in 1902. Making use of a short-lived radioactive isotope, they traced the variation of the intensity of radiation with time. They found the intensity fell off with time in what is called an "exponential manner."

This can be true only if individual radioactive atoms break down at a rate that is a fixed fraction of the total number of such atoms present. This fraction, let us say, is 0.02 of the atoms present per second. If we begin with 1,000,000,000,000 atoms; then in the first second 20,000,000,000 atoms will break down. We can't tell which twenty billion will break down, of course. If we were considering a particular atom, we couldn't tell whether that one atom would break down in the first second or after five seconds or after five years.

This is quite analogous to a much more familiar situation. Given a million Americans aged 35, insurance companies (from a thorough study of statistics) can predict with reasonable accuracy how many of them will die in the course of the next year, assuming the year to be a "normal" one. They could not point out individual Americans who will die in that year, or predict the particular year of death of a particular American. They can only make general predictions where large numbers of faceless individuals are concerned. Where insurance men work with millions of human beings, physicists work with trillions of trillions of atoms, and the predictions of the latter are correspondingly more accurate.

Radioactive breakdown involves a fixed breakdown rate as time goes on. Let us suppose that this was true of human deaths. Let us suppose that out of 1,000,000 Americans aged 35 some 0.2 percent—that is 2000—will die in the course of the year. At the end of the year 998,000 men are left. If now 0.2 percent of those die in the next year, 1996 will die, leaving 996,004. In the third year, keeping the rate constant, 1992 will die, leaving 994,012, and so on.

The number of men dying would, in such a case, decrease evenly with the number of men still alive. There would be no particular year in which you could predict that the last man would die, for in any given year only a small percentage of those alive would die. Naturally, this analysis would only be reasonably correct as long as the number of men remained great enough for statistical methods to be reasonably accurate. However, if you started with an extremely large number of men, you would expect some of them to live hundreds of thousands of years.

This does not actually happen because the human death rate does not remain constant as the men grow older. It rises steadily, and very old men have a very high death rate. For that reason, no matter how large a number of 35-year-old men you begin with, all will be dead in less than a century.

In the case of radioactive atoms, the "death rate" through breakdown does not change with time, and while some atoms break down almost at once, other atoms of the same kind may refrain from breaking down for indefinitely long periods of time. One cannot therefore speak of the "lifetime" of a radioactive atom since that "lifetime" can be anything.

It is characteristic of such a "fixed death rate" situation, however, that for a given value of that death rate there is a specific interval of time during which half the original atoms would break down. This specific interval of time, named the *half-life* by Rutherford in 1904, would remain the same, however large or

Half-life

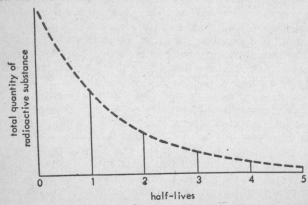

total quantity of radioactive substance

half-lives

however small (within statistical reason) the original number of atoms. For given isotopes, such a half-life was found to be virtually independent of environmental conditions such as temperature and pressure. Physicists have found ways of imposing minor changes— a few percent at most—on the half-lives of a few specific types of radioactive atoms, but such cases are quite exceptional.

Let us say that the half-life of a particular isotope is one year. This means that if you begin with two trillion atoms of that isotope, you will have only one trillion left at the end of the year. With the number of atoms present having declined to half, the number of breakdowns also declines to half, and only half a trillion vanish in the next year, leaving half a trillion. After a third year, a quarter-trillion would be left, and so on.

To generalize: Given any number of radioactive atoms, half will break down during the first half-life period, half of those left will break down during the second half-life period, and so on indefinitely—or at least until the total number of atoms involved becomes small enough for statistical methods no longer to apply with reasonable accuracy.

To know the half-life of an isotope, then, is to know in capsule form how many breakdowns will take place in a given quantity in a given time, and therefore, how intensely radioactive the isotope is. You can also trace what the intensity will be at any time in the future and what it was at any time in the past.

The half-lives of radioactive isotopes vary in length from the vanishingly small to the immensely long. In the intermediate range it is possible to determine half-lives directly from observed breakdown rates. For instance, the half-life of radium-226 is 1620 years.

To find half-lives much longer than this, indirect methods may be used. Consider the case of uranium-238, for instance. In any given sample of uranium ore, the uranium atoms are breaking down, but at so small a rate that we can safely assume that over some limited period of time, the number of uranium atoms present is virtually constant. Call that number, N_u. A given fraction (F_u) of the number of uranium atoms present breaks down each second. The total number of uranium atoms breaking down each second is therefore $F_u N_u$.

In the course of its breakdown, uranium-238 forms radium-226. It doesn't do this directly, for there are four other radioactive isotopes between, but this can be shown not to matter. We can legitimately simplify matters for the present by assuming that radium-226 is formed directly from uranium-238. Since $F_u N_u$ uranium-238 atoms are breaking down each second, $F_u N_u$ radium-226 atoms are being formed each second.

As radium-226 is formed, it also begins to break down at a rate that is a fixed fraction of the number of radium-226 atoms present, $F_r N_r$. As the radium-226 atoms form and accumulate from uranium-238, the number of radium-226 atoms breaking down increases until the point is reached where the number breaking down is equal to the number being formed. At that point, the number of radium-226 atoms actually present reaches a constant value, and radium-226 is in *radioactive equilibrium* with uranium-238. At radioactive equilibrium:

$$F_u N_u = F_r N_r \qquad \text{(Equation 8–8)}$$

or, rearranging:

$$\frac{F_u}{F_r} = \frac{N_r}{N_u} \qquad \text{(Equation 8–9)}$$

It can be shown that the fraction of a particular isotope breaking down each second is inversely proportional to the half-life of that particular isotope. The longer the half-life, the smaller the fraction of atoms present breaking down in one second. If the half-life of uranium-238 is symbolized as H_u and that of radium-226 as H_r, we can say:

$$\frac{F_u}{F_r} = \frac{H_r}{H_u} \qquad \text{(Equation 8–10)}$$

Combining Equations 8–9 and 8–10, we have:

$$\frac{H_r}{H_u} = \frac{N_r}{N_u} \qquad \text{(Equation 8–11)}$$

At radioactive equilibrium, in other words, the ratio of the quantity of parent and daughter atoms present is equal to the ratio of their half-lives. In uranium ores, there are 2,800,000 times as many uranium-238 atoms present as radium-226 atoms. The half-life of uranium-238 must be 2,800,000 times as long as that of radium-226, or just about 4,500,000,000 years.

It is not surprising then, that uranium-238 still exists in the earth's crust. If the solar system is five to six billion years old (as is now believed), then there has been time for only little more than half the uranium-238 originally present to have broken down.

The half-life of uranium-235 is shorter than that of uranium-

238; it is only 713,000,000 years. This is still long enough for something like one percent of the original amount present at the time of the origin of the solar system to remain in existence today. However, it is not surprising that only seven out of a thousand uranium atoms now existing are uranium-235.

Any radioactive isotope with a half-life of less than 500,000,000 years would not be present on earth today in more than vanishingly small quantities, unless it were formed out of a longer-lived ancestor. In the $4x + 2$ series, only uranium-238 qualifies for existence, and in the $4x + 3$ series, only uranium-235 does.

The only $4x + 0$ atom with a long enough half-life to exist today and give rise to a radioactive series is, of course, thorium-232. Its half-life is no less than 13,900,000,000 years.

Indirect methods can also be used to determine very short half-lives. For instance, it has been found that among those radioactive isotopes that emit alpha particles, the energy of the particles is inversely proportional to the half-life, in a moderately complicated fashion. Therefore, the half-life can be calculated from the energy of the alpha particles (which can be determined by noting how far they will penetrate a given type of substance). The half-life of polonium-212, for example, is found to be 0.0000003 seconds.

If isotopes of a given element differ among themselves inappreciably in chemical properties, they differ among themselves enormously in nuclear properties such as half-life. Thorium-232, as stated above, has a half-life of nearly fourteen billion years, but thorium-231 (differing in the lack of a single neutron in the nucleus) has a half-life of just about one day!

Stable Isotopes

If we go over the three radioactive series presented in Tables V, VI, and VII, we see that they include radioactive isotopes of elements ordinarily considered stable. There are five such isotopes of bismuth, with atomic weights of 210, 211, 212, 214 and 215; four of thallium, with atomic weights of 206, 207, 208 and 210; and four of lead, with atomic weights of 210, 211, 212 and 214. Each of these must possess at least one stable isotope as well, since each is found in the soil in appreciable quantities and in nonradioactive form. Indeed, the radioactive series include three different isotopes of lead—206, 207 and 208—each one of which is stable.

Nevertheless, all these isotopes, stable as well as unstable, are involved with radioactivity. It is fair to ask if elements that are in no way involved with radioactivity may nevertheless consist of two or more isotopes. If so, the fact would be difficult to establish, since ordinary laboratory methods do not suffice to separate isotopes (save in exceptional cases, see page 143) and since radioactivity cannot be relied on to help.

But suppose the atoms of an element are ionized, as in the formation of positive rays (see page 112). The atoms, each with an electron removed, would all have an identical charge of $+1$. If the element consisted of two or more isotopes, however, the ions would fall into groups that differ in mass.

Suppose, now, a stream of these positive ions are made to pass through a magnetic field. Their path is bound to curve and the extent of the curvature would depend upon the charge and mass of the individual particles. The charge would be the same in every case, but the mass would not be. The more massive ions curve less sharply than do the less massive ones. If the stream of positive rays is allowed to fall on a photographic plate, they would form one spot if all the ions were alike in mass, but more than one spot if the ions formed groups of different mass. Furthermore, if the groups were unequal in size, the larger group would form a larger, darker spot.

In 1912, J. J. Thomson, the discoverer of the electron, performed an experiment of this sort with neon. Positive rays formed of neon ions made two spots on the plate, corresponding to what would be expected for neon-20 and neon-22. The former spot was some ten times as large as the latter; from this is could be concluded that neon consisted of two stable isotopes, neon-20 and neon-22, in a ratio of about 10 to 1. (Eventually, it was discovered that a third stable isotope, neon-21, existed in very small quantities, and that in every 1000 neon atoms, 909 were neon-22, 88 were neon-22 and 3 were neon-21.)

In 1919, the English physicist Francis William Aston (1877–1945), who had worked with Thomson on this problem, constructed an improved device for analyzing positive rays. In his device, the positive rays consisting of ions of a given mass did not simply form a smear on the photographic plate. They were curved in such a way as to focus at a point, thus allowing for finer resolution. As a result, the beam of ions produced from a given element was spread out into a succession of points (a "mass spectrum" rather than a light spectrum). From the position of the points one could deduce the mass of the individual isotopes, and from their darkness, the frequency (or *relative abundance*) with which each occurred in the element. The instrument was termed a *mass spectrograph*.

The use of the mass spectrograph made it quite plain that most stable elements consisted of two or more stable isotopes. A complete list of such stable isotopes* is given in Table VIII.

A number of points can be made concerning Table VIII. In the first place, although most of the 81 stable elements consist of two or more stable isotopes (with tin made up of no less than ten), there remain no less than 20 elements consisting of but a single isotope. (Indeed, two elements of atomic number less than 84 possess no stable isotopes at all. These two, with atomic numbers 43 and 61, will be discussed on page 175.)

Properly speaking, one ought not speak of "a single isotope," since isotopes were originally defined as two or more kinds of atom falling in the same place in the periodic table. One might as well speak of "one twin." For that reason one sometimes speaks of *nuclides,* meaning an atom with a characteristic nuclear structure. One can certainly speak of one nuclide. However, the term "isotope" is so well-established that I will continue to speak of "a single isotope" with the assurance that I will be correctly understood.

Not all the 282 nuclides listed in Table VIII are indeed completely stable. Some eighteen of them, it turns out, are radioactive, though always with such extended half-lives that the radioactivity they display is feeble indeed. Some, with half-lives in the quadrillions of years, have radioactivities so weak that they may be ignored for all practical purposes. Seven, however, are perceptibly radioactive, and these are included in Table IX.

* Or isotopes that are so feebly radioactive that they may just about be considered stable.

TABLE VIII—*The Stable Isotopes*

Atomic Number	Element	Isotope Weights
1	Hydrogen	1, 2
2	Helium	3, 4
3	Lithium	6, 7
4	Beryllium	9
5	Boron	10, 11
6	Carbon	12, 13
7	Nitrogen	14, 15

The Stable Isotopes (continued)

Atomic Number	Element	Isotope Weights
8	Oxygen	16, 17, 18
9	Fluorine	19
10	Neon	20, 21, 22
11	Sodium	23
12	Magnesium	24, 25, 26
13	Aluminum	27
14	Silicon	28, 29, 30
15	Phosphorus	31
16	Sulfur	32, 33, 34, 36
17	Chlorine	35, 37
18	Argon	36, 38, 40
19	Potassium	39, 40, 41
20	Calcium	40, 42, 43, 44, 46, 48
21	Scandium	45
22	Titanium	46, 47, 48, 49, 50
23	Vanadium	50, 51
24	Chromium	50, 52, 53, 54
25	Manganese	55
26	Iron	54, 56, 57, 58
27	Cobalt	59
28	Nickel	58, 60, 61, 62, 64
29	Copper	63, 65
30	Zinc	64, 66, 67, 68, 70
31	Gallium	69, 71
32	Germanium	70, 72, 73, 74, 76
33	Arsenic	75
34	Selenium	74, 76, 77, 78, 80, 82
35	Bromine	79, 81
36	Krypton	78, 80, 82, 83, 84, 86
37	Rubidium	85, 87
38	Strontium	84, 86, 87, 88
39	Yttrium	89
40	Zirconium	90, 91, 92, 94, 96
41	Niobium	93
42	Molybdenum	92, 94, 95, 96, 97, 98
44	Ruthenium	96, 98, 99, 100, 101, 102, 104
45	Rhodium	103
46	Palladium	102, 104, 105, 106, 108, 110
47	Silver	107, 109

The Stable Isotopes (continued)

Atomic Number	Element	Isotope Weights
48	Cadmium	106, 108, 110, 111, 112, 113, 114, 116
49	Indium	113, 115
50	Tin	112, 114, 115, 116, 117, 118, 119, 120, 122, 124
51	Antimony	121, 123
52	Tellurium	120, 122, 123, 124, 125, 126, 128, 130
53	Iodine	127
54	Xenon	124, 126, 128, 129, 130, 131, 132, 134, 136
55	Cesium	133
56	Barium	130, 132, 134, 135, 136, 137, 138
57	Lanthanum	138, 139
58	Cerium	136, 138, 140, 142
59	Praseodymium	141
60	Neodymium	142, 143, 144, 145, 146, 148, 150
62	Samarium	144, 147, 148, 149, 150, 152, 154
63	Europium	151, 153
64	Gadolinium	152, 154, 155, 156, 157, 158, 160
65	Terbium	159
66	Dysprosium	156, 158, 160, 161, 162, 163, 164
67	Holmium	165
68	Erbium	162, 164, 166, 167, 168, 170
69	Thulium	169
70	Ytterbium	168, 170, 171, 172, 173, 174, 176
71	Lutetium	175, 176
72	Hafnium	174, 176, 177, 178, 179, 180
73	Tantalum	180, 181
74	Tungsten	180, 182, 183, 184, 186
75	Rhenium	185, 187
76	Osmium	184, 186, 187, 188, 189, 190, 192

142   *Understanding Physics*

The Stable Isotopes (continued)

Atomic Number	Element	Isotope Weights
77	Iridium	191, 193
78	Platinum	190, 192, 194, 195, 196, 198
79	Gold	197
80	Mercury	196, 198, 199, 200, 201, 202, 204
81	Thallium	203, 205
82	Lead	204, 206, 207, 208
83	Bismuth	209

It might seem surprising that this radioactivity among the lighter elements was not detected sooner than it was—particularly in the case of potassium-40. Potassium is a very common element, and potassium-40 (alone among the isotopes listed in Table IX) has a shorter half-life and is therefore more intensely radioactive than is either uranium-238 or thorium-232.

The answer to that is twofold. In the first place, potassium-40 makes up only one atom out of every 10,000 in potassium, so that it is not as common as it seems. In the second place, uranium and thorium are the parents of a series of intensely radioactive isotopes. It is the daughter atoms, rather than uranium or thorium themselves, that gave rise to the effects observed by Becquerel and the Curies.

None of the long-lived radioactive isotopes of the lighter elements serve as parents for a radioactive series. In every case they emit beta particles and are, in a single step, converted into a stable isotope of the element one atomic number higher. Thus, rubidium-87 becomes stable strontium-87, lanthanum-138 becomes stable cerium-138, and so on.

Potassium-40 introduces a slight variation. Of all the potassium-40 atoms that break down, some eighty-nine percent do indeed emit a beta particle and become stable calcium-40. The remaining eleven percent absorb an electron into the nucleus. This electron is taken from the innermost extranuclear shell, the K-shell (see page 66), and the process is therefore known as *K-capture*. An electron taken into the nucleus serves to cancel the positive charge of one proton and produce an additional neutron. The total number of nucleons is not changed and neither, therefore, is the atomic weight. The atomic number, however, decreases by one. By K-capture, potassium-40 (atomic number 19) becomes the stable argon-40 (atomic number 18).

In some ways, the most remarkable of the stable isotopes is hydrogen-2, the nucleus of which is made up of one proton and one neutron, instead of one proton only as in hydrogen-1. The mass ratio between the two stable isotopes of hydrogen is much greater than is true of any two stable isotopes of any other element.

Thus, uranium-238 is 238/235, or 1.013 times the mass of uranium-235. Tin-124, the heaviest of the stable isotopes of that element, is 1.107 times the mass of tin-112, the lightest. Oxygen-18 is 1.125 times the mass of oxygen-16. But compare that with hydrogen-2, which is 2.000 times the mass of hydrogen-1.

This great difference in relative masses between the two hydrogen isotopes means that the two differ considerably more in their physical and chemical properties than isotopes generally do. The boiling point of ordinary hydrogen is 20.38°K, whereas hydrogen made up of hydrogen-2 only ("heavy hydrogen") has a boiling point of 23.50°K.

Again, ordinary water has a density of 1.000 gram per cubic centimeter and a freezing point of 273.1°K (0°C). Water with molecules containing hydrogen-2 only ("heavy water") has a density of 1.108 grams per cubic centimeter and a freezing point of 276.9°K (3.8°C).

So marked are the differences between hydrogen-1 and hydrogen-2 that the latter is given the special name of *deuterium* (from a Greek word for "second"). Its symbol is D and heavy hydrogen can be written D_2, while heavy water can be written D_2O.

In the early days of isotope work, physicists had suspected the existence of deuterium because the atomic weight of hydrogen seemed slightly higher than it ought to be. The single electron, it

TABLE IX—*Lighter Radioactive Nuclides*

Nuclide	Half-Life (years)
Potassium-40	1,300,000,000
Rubidium-87	47,000,000,000
Lanthanum-138	110,000,000,000
Samarium-146	106,000,000,000
Lutetium-176	36,000,000,000
Rhenium-187	70,000,000,000
Platinum-190	700,000,000,000

was calculated, would have its energy levels somewhat differently distributed in hydrogen-2 than in hydrogen-1, so faint lines of the former ought to appear near the heavy lines of the latter in the hydrogen spectrum. This was not observed, nor was hydrogen-2 located by mass spectrograph. One reason for this is that hydrogen-2 is quite rare; only one atom out of 7000 in ordinary hydrogen is hydrogen-2.

The American chemist Harold Clayton Urey (1893–   ) began, in 1931, with four liters of liquid hydrogen and let it slowly evaporate to one cubic centimeter. He reasoned that hydrogen-2 would evaporate more slowly and would be concentrated in the final bit. He was right. When he studied the spectrum of that last residue, he detected the lines of deuterium precisely where calculations had predicted they would be.

...

CHAPTER 9

Nuclear Chemistry

Mass Number

One might try, out of a spirit of neatness, to divide the atom
cleanly between the two major physical sciences, awarding the
electrons to the chemist and the nucleus to the physicist.

To attempt such a cleancut division would, however, be a
violation of the spirit of science, which is all-one-piece. The structure of the nucleus, however remote that might seem from the
world of ordinary chemical reactions, must nevertheless be of keen
interest to the chemist, if only because of its effect upon that
fundamental chemical datum, the atomic weight.

By the time the nineteenth century had come to its end, the
matter of the atomic weight seemed settled. Each element had a
characteristic atomic weight, chemists believed, and the only
future in that respect was the ever greater precision with which
the fourth and fifth decimal places might be determined.

Then came the discovery of isotopes and all that had seemed
certain about atomic weights immediately went into discard. The
notion, dating from Dalton, that all atoms of a single element
possessed identical mass and that the atomic weight expressed
this mass was seen to be false. Instead, most elements were made
up of two or more varieties of atoms differing in mass. The atomic
weight was merely the weighted average of the masses of these
isotopes.

145

If the term "atomic weight" is reserved for this weighted average of the isotope masses as found in their natural distribution within an element, then one ought not to speak of the atomic weight of an individual isotope (as I have been doing so far in this book). It is better to use a different phrase and speak of the relative mass of an individual isotope as its *mass number*.

We can say, therefore, that neon is made up of three isotopes of mass numbers 20, 21, and 22. Neon-20 makes up about nine-tenths of all the neon atoms, while neon-22 makes up most of the remaining tenth. We can neglect neon-21 as occurring in too small a concentration to affect the result materially, and content ourselves by taking the average of ten atoms of neon, nine of which have a mass of 20 and one of which has a mass of 22. We arrive at a result of 20.2, which is roughly the atomic weight of neon.

Again, chlorine is made up of two isotopes of mass numbers 35 and 37, with chlorine-35 making up three-quarters of the whole, and chlorine-37 making up the remaining quarter. If we average the mass of four atoms, three of which have a mass of 35 and one of which has a mass of 37, we end with a result of 35.5, which is also roughly the atomic weight of chlorine.

All the nineteenth century demonstrations that Prout's hypothesis (see page 24) was false—because the atomic weights of the various elements were not necessarily integral multiples of the atomic weight of hydrogen—were shown to be irrelevant. The mass numbers of the various isotopes were, without exception, all found to be very nearly exact multiples of the mass of the hydrogen atom, and Prout's hypothesis was re-established in a more sophisticated form. The various elements were not built up out of hydrogen atoms exactly, but (ignoring the almost massless electrons) they were built up out of nucleons of nearly identical mass, while the hydrogen atom itself is built up out of a single nucleon.

Atomic weights that are nearly whole numbers in value are so because the particular element is made up of a single isotope, as in the case of aluminum, or of two or more isotopes, with one vastly predominant in relative abundance. An example of the latter situation is calcium, which is made up of six stable isotopes with mass numbers of 40, 42, 43, 44, 46, and 48, but with calcium-40 making up ninety-seven percent of the whole. It is because so many of the lighter elements fall into one of these two classes that Prout found reason to advance his hypothesis in the first place.

It is the imbalance of isotopes that causes some elements to be "out of order" in the periodic table. Thus, cobalt, with an atomic number of 27, consists of the single isotope of mass number 59. Its atomic weight, therefore, is about 58.9.* We would expect nickel, with the higher atomic number 28, to have a higher atomic weight as well. Nickel consists of five isotopes with mass numbers 58, 60, 61, 62, and 64; and, not surprisingly, four of those isotopes have mass numbers higher than the single cobalt isotope. However, it is the lightest of the nickel isotopes, nickel-58, which happens to be predominant. There are twice as many nickel-58 atoms than all the other nickel atoms put together. The atomic weight of nickel is thus pulled down to 58.7, which is somewhat less than that of cobalt.

The atomic weight is thus deprived of its fundamental character, and it is not truly characteristic of an element. What made it seem characteristic was the fact that the various isotopes of an element have virtually identical properties. The processes that led to the concentration of the compounds of an element in various places in the earth's structure, or to the isolation of the element in the laboratory, affected all the isotopes alike. Each sample of an element, however produced, would therefore contain the various isotopes in virtually identical proportions and would therefore display the same, apparently characteristic, atomic weight.

Yet there are exceptional cases, and the most dramatic is that of lead. Each of the radioactive series (see Chapter 8) ends in a particular lead isotope. The two series that begin with uranium isotopes as parent atoms produce lead-206 and lead-207, with lead-206 far in the lead since there is so much more uranium-238 than uranium-235. As for the thorium series, that ends in uranium-208.

The atomic weight of ordinary lead, found in nonradioactive ores, is about 207.2. In uranium ores, with lead-206 having been produced steadily over geologic periods, the atomic weight should be distinctly less, while in thorium ores it should be distinctly more. In 1914, the American chemist Theodore William Richards (1868–1928) carried through atomic-weight determinations and found, indeed, that lead obtained from uranium ores had atomic weights that ran as low as 206.1. Lead from thorium ores gave atomic weights as high as 207.9.

Where radioactivity is not involved, such large variations are not to be expected. Still, the atomic weights of some of the lighter elements were found to vary slightly in accordance with the con-

* Why not exactly 59? See page 185.

ditions under which the element was produced. For instance, the relative distribution of oxygen-16 and oxygen-18 in the calcium carbonate ($CaCO_3$) of seashells has been shown to depend on the temperature of the water in which the organism that formed the seashell was living. Delicate measurements of isotope ratios in fossil seashells have therefore been used to determine the temperature of ocean water at different periods of earth's geologic past.

The existence of oxygen isotopes introduced a particular difficulty in connection with atomic weights. By a convention as old as Berzelius, the atomic weights had been determined on a standard that set the atomic weight of oxygen arbitrarily equal to 16.0000. In 1929, however, the American chemist William Francis Giauque (1895–   ) showed that oxygen consisted of three isotopes—oxygen-16, oxygen-17 and oxygen-18—and that the atomic weight of oxygen had therefore to represent the weighted average of three mass numbers.

To be sure, oxygen-16 was by far the most common of the three, making up 99.759 percent of the whole, so that one wasn't very far off in pretending that oxygen was made up of a single isotope. For a generation after the discovery, therefore, chemists tended to ignore the oxygen isotopes and continue the atomic weights on the old basis. Such atomic weights came to be called *chemical atomic weights*.

Physicists, however, preferred to set the mass of the oxygen-16 isotope at 16.0000 and to determine all other masses on that basis. Their reasoning was that the mass number of an isotope was characteristic and unalterable, whereas the atomic weight of a multi-isotope element would alter with changes in relative abundance of those isotopes from sample to sample.

On the basis of oxygen-16 = 16.0000, a new list of atomic weights, the *physical atomic weights*, was drawn up. The atomic weight of oxygen on this new basis was 16.0044 (oxygen-17 and oxygen-18 pulling up the average), and this is 0.027 percent higher than its chemical atomic weight of 16.0000. This same difference would exist throughout the entire list of elements, and while this difference is small, it acts as an unnecessary source of confusion in refined work.

In 1961, physicists and chemists reached a compromise. It was agreed to determine atomic weights on the basis of allowing the carbon-12 isotope to have a mass of 12.0000. As the physicists desired, this tied the atomic weights to a fixed and characteristic mass number. In addition, an isotope was chosen for the purpose which would produce a set of atomic weights as nearly identical to the old chemical atomic weights as possible. Thus the atomic weight of oxygen under this new system is 15.9994, which is only 0.0037 percent less than the chemical atomic weight. The atomic weights given in Table II of Chapter 1, are on a carbon-12 = 12.0000 basis.

The atomic weight, since it is the weighted average of the mass numbers of the naturally-occurring isotopes, can truly be applied only to those elements that are primordial; that is, that have been in the earth from the beginning, when, presumably, the different isotopes made their appearance at the same time. This includes only 83 elements altogether. There are first the 81 stable elements (with atomic numbers from 1 to 83 inclusive, minus atomic numbers 43 and 61), and then the nearly stable elements thorium and uranium.

The elements that appear in the earth only because they are formed from uranium or thorium appear in the form of isotopes of different mass number, depending on whether they are found in uranium or thorium ore. One cannot form a true average and obtain a real atomic number. It is therefore usual, in the case of these elements (and of other unstable elements to be considered in the next chapter), to use the mass number of the most long-lived known isotope as a kind of substitute atomic weight. In tables of atomic weights, these mass numbers are usually included in brackets, as in Table X. Of the isotopes which appear in this table, those of radon and radium appear naturally in the uranium series, while those of francium, actinium and protactinium appear naturally in the actinium series. Polonium-209 and astatine-210 do not occur naturally at all but have been artificially produced (see page 175).

TABLE X—*"Atomic Weight" of Radioactive Elements*

Element	Mass Number of Most Long-Lived Isotope	Half-Life	
84—Polonium	[209]	103	years
85—Astatine	[210]	8.3	hours
86—Radon	[222]	3.8	days
87—Francium	[223]	22	minutes
88—Radium	[226]	1,620	years
89—Actinium	[227]	21.2	years
91—Protactinium	[231]	32,480	years

Radioactive Dating

The lead isotopes played a role not only in re-orienting the chemical view of atomic weights, but also of the geological view of terrestrial history.

As long ago as 1907, the American physicist Bertram Borden Boltwood (1870–1927) suggested that radioactive series could be used as a method for determining the age of minerals.

Suppose a particular layer of rock containing uranium or thorium was laid down at a certain time in the past as a solid by sedimentation from the sea or by freezing from a volcanic melt. Once such a solid had made its appearance, the uranium or thorium atoms within it would be "trapped." When some broke down and eventually formed lead atoms, those atoms would be trapped, too.

During the entire stretch of time that would have elapsed since the layer had solidified, the uranium or thorium would be breaking down and the lead content would, in consequence, be rising. It would seem, then, that the uranium/lead and thorium/lead ratios within solid rocks would increase steadily with time.

Since Rutherford had already worked out the concept of half-life, it seemed furthermore that this increase would take place at a known rate, so that from the uranium/lead or thorium/lead ratio at any instant of time (the present, for instance), the time lapse since the rock had solidified could be calculated. Since the half-life of both uranium-238 and thorium-232 is so immensely long, time lapses of billions of years could be calculated with assurance.

A possible difficulty rests in the fact that one cannot be sure that the lead in such rocks was produced entirely through uranium or thorium breakdown. An indefinite amount might be primordial and might have been trapped along with uranium in the rock at the moment of its solidification. Such lead would obviously have no connection with the uranium and would seriously confuse the issue.

This difficulty was resolved when the mass spectrograph made it possible to determine the relative abundance of the isotopes in lead found in nonradioactive rocks. Such lead contained four stable isotopes, with mass numbers 204, 206, 207, and 208; and lead-204 made up 1.48 percent (1/67.5) of the whole.

This is fortunate, for lead-204 is not produced as the end product of any radioactive series, and its occurrence is not affected by radioactivity. If in the lead content of radioactive rocks the lead-204 concentration is determined and multiplied by 67.5, then the total quantity of primordial lead can be determined. Any lead over and above this quantity would have been produced by radioactive breakdown.

By using the uranium/lead ratio, and allowing for the presence of lead-204, rocks have been found which have been solid for over 4,000,000,000 years. This is taken as the best evidence yet obtained for the extreme age of the earth.

Uranium and thorium are not, of course, among the more common elements, and rocks containing sufficient uranium and thorium to make reasonably reliable age determinations of this sort are found only in restricted areas. However, use can also be made of the long-lived radioactive isotopes rubidium-87 and potassium-40, each of which is much more widely distributed than is either uranium or thorium. In the case of rubidium-87, one can determine the rubidium/strontium ratios in rocks, since rubidium-87 decays to the stable strontium-87. (Primordial strontium can be estimated by noting the quantity of other stable strontium isotopes present, these others not being formed in radioactive breakdown.) Rubidium-containing minerals that have been solid for nearly 4,000,000,000 years have been located.

Potassium-40 offers an interesting situation. Mostly, it breaks down to calcium-40; but calcium-40 is very common in the earth's crust, and it is impractical to try to distinguish "radiogenic" calcium-40 (that which has arisen through radioactive breakdown) from primordial calcium-40. However, a fixed proportion of the potassium-40 atoms break down by K-capture (see page 142) to form argon-40.

Argon is one of the inert gases found in the atmosphere. All the isotopes of the various inert gases, with the single exception of argon-40, are present in almost vanishingly small quantities. This situation probably reflects a time in earth's early history when its mass was too small or its temperature too high to retain any of the gaseous elements except in the form of solid compounds. Since the inert gases do not form any compounds to speak of, they were all lost.

Argon-40, however, occurs in quantity, making up about one percent of the atmosphere. It seems likely that all this argon-40 was only formed after the earth had attained its present mass and temperature (at which time it could retain the heavier inert gases), and that it had been formed, presumably, from potassium-40. If one calculates the time it would have taken the present quantity of argon-40 to have accumulated from scratch, it would appear that the earth has existed in approximately its present form for 4,000,000,000 years.

Since a variety of methods independently agree, the question of the earth's age is taken as settled—or at least (remember the fate of many previous "settled" points) it is so taken until further notice.

Nuclear Reactions

As long as it was thought that an atom was a structureless, indivisible particle, it seemed an inevitable consequence that its nature could not be altered in the laboratory. However, once the atom was found to consist of numerous subatomic particles in a characteristic arrangement, the thought at once arose that this arrangement might somehow be altered.

The outer electrons of an atom may have their arrangement altered easily enough. Collisions among atoms and molecules with forces to be expected at temperatures attainable in nineteenth century laboratories sufficed for the purpose. It was these electron rearrangements that produced the familiar chemical reactions that were the established province of the chemist.

But what about rearrangements among the particles within the nucleus? These would alter the very fundamental nature of an atom and convert one element into another.

To smash atoms together so hard that the outer cushion of electrons is smashed through and nucleus meets nucleus requires extraordinarily high temperatures. Fortunately, as the twentieth century opened there was an obvious way of bypassing the need for such temperatures. Radioactive elements furnished a supply of subatomic particles at room temperature. One of them, the alpha particle, was a bare atomic nucleus (that of helium). Alpha particles are emitted with enough energy to smash through the electron barrier and, if properly aimed, strike the nucleus of a target atom.

It is impossible, of course, to aim an alpha particle at a given nucleus, but, statistically speaking, if enough alpha particles are fired, some will strike nuclei. It was through such collisions and near-collisions that Rutherford worked out the concept of the nuclear atom and estimated the size of the nucleus (see page 58).

Still, a collision that merely results in a deflection or a bounce alters the nature of neither the target nucleus nor the alpha particle. Something more is needed, and in a series of experiments, the results of which he described in 1919, Rutherford worked out the necessary evidence that something more is occasionally obtained. He began by placing a source of energetic alpha particles inside a closed cylinder, one end of which was coated with a layer of zinc sulfide.

Now, whenever an alpha particle strikes the zinc sulfide, it gives rise to a tiny flash of luminescence, or scintillation. This arises because the kinetic energy of the particle excites the zinc sulfide molecule which, in returning to its ground state, emits a photon of visible light. (This phenomenon was first observed by Becquerel in 1899 and was later put to use in the preparation of luminescent objects. Tiny quantities of radium compounds mixed with zinc sulfide or some other appropriate substance would produce light flashes that would be clearly visible in the dark. In the 1920's there grew up quite a fad for watches with numerals marked out with such luminescent materials.)

When scintillating zinc sulfide screens were viewed under some magnification in the dark (with eyes well-accustomed to darkness and therefore particularly sensitive to feeble light) individual scintillations could be made out and therefore, surprisingly enough, so could the effect of single alpha particles. By counting the number of flashes over a given area in a given time, one could estimate the total number of disintegrations per second in a known mass of radioactive substance, and from this (for instance) one could calculate the half-life. Rutherford in his experiments was making use of what is now known as a *scintillation counter*.

Modern scintillation counters make use of more efficient scintillators, of phototubes to detect the flashes, and of appropriate electronic circuits to count them.

The number of scintillations produced by a given alpha particle source is reduced if a gas such as oxygen or carbon dioxide is introduced into the tube. Through collision and deflection, the gas slows the alpha particles to the point where some pick up electrons and become ordinary helium atoms. Those that manage to reach the screeen are fewer and less energetic.

If hydrogen is introduced into the tube, however, particularly bright scintillations suddenly appear. These can best be interpreted by supposing that occasionally an alpha particle will strike the nucleus of a hydrogen nucleus (a single proton) squarely and send it hurtling forward, away from what had been its associated electron. In this fashion, the bare proton can be made to move far more rapidly than could the massive nuclei of carbon and oxygen. In fact, the proton moves quickly enough to strike the screen with sufficient force to produce the unusually bright scintillations.

Rutherford found that when nitrogen was introduced into the tube, what looked like proton scintillations appeared. The nitrogen nucleus itself could not be forcibly hurled forward any more than could those of carbon or oxygen, but perhaps a proton had been knocked out of the nitrogen nucleus by the alpha particle.

This was confirmed in 1925 by the English physicist Patrick Maynard Stuart Blackett (1897–   ), who allowed the alpha particle bombardment of nitrogen to proceed in a Wilson cloud chamber. The alpha particle usually made a straight streak of water droplets without striking any nucleus, disappearing when its energy had been sufficiently nibbled away for the particle to pick up electrons and become an atom. Once in every 50,000 cases or so, however, there was a collision.

The alpha particle streak therefore ended in a fork. One side of the fork was long and thinner than the original track; it was the proton, carrying a smaller charge (+ 1, rather than the alpha particle's + 2) and producing fewer ionizations. The other side of the fork was thick and short. It was the recoiling nitrogen nucleus from which numerous electrons had been stripped, and its high positive charge made it an efficient ionizer. However, it moved rather slowly, quickly picked up electrons once more, and, neutral again, ceased ionizing. There was no sign of the alpha particle after the collision, so it must have joined the nitrogen nucleus.

In the light of all this, it was not difficult to see that Rutherford, in 1919, had produced the first case of a deliberate rearrangement of nuclear structure through human efforts. It was the first man-made *nuclear reaction*. (In a sense, this is a kind of "nuclear chemistry," for the nucleons were being shuffled about in fashion analogous to the shuffling of electrons in ordinary chemistry.)

Suppose we begin with a nitrogen nucleus (seven protons and seven neutrons), add to it an alpha particle (two protons and two neutrons) and subtract the single proton that is knocked out. What is left then is an atom of eight protons and nine neutrons, which is oxygen-17. We can therefore write:

$$_7N^{14} + {}_2He^4 \longrightarrow {}_1H^1 + {}_xO^{17}$$ (Equation 9–1)

where the subscripts are atomic numbers and the superscripts mass numbers. The $_2He^4$ represents the helium nucleus, or alpha particle, while the $_1H^1$ is the hydrogen nucleus, or proton. Notice that the atomic numbers add up to 9 on either side of the arrow and the mass numbers add up to 18. Such a balance must be preserved in all nuclear reactions if the laws of conservation of electric charge and of mass are to be preserved.

Physicists have devised briefer methods of writing such nuclear reactions. The atomic number is omitted since the name of the element fixes that number. The alpha particle is symbolized as a, and the proton as p. The nuclear reaction given in Equation 9–1 can then be written as: $N^{14}(a,p)O^{17}$.

According to this system we have the target nucleus at the far left, then, in the parentheses, first the nature of the particle striking the target and then the particle knocked out of the target. Finally, on the extreme right, is the residual nucleus. The usefulness of this system, quite apart from its conciseness, is that it makes it easy to speak of a whole family of (a,p) nuclear reactions. In all such reactions, the residual nucleus is one higher in atomic number and three higher in mass number than the target nucleus.

Other (a,p) reactions were brought about by Rutherford, but there is a limit to what can be done in this direction. Both the alpha particle and the target nucleus are positively charged and repel each other. This repulsion increases with the atomic number of the nucleus, and for nuclei of elements beyond potassium (with a charge of $+19$) the repulsion is so strong that even the most energetic alpha particles produced by radioactive atoms lack the energy required to overcome that repulsion and strike the nucleus.

The search was on, therefore, for methods of obtaining sub-atomic particles with energies greater than those encountered in radioactivity.

The Electron-Volt

A charged particle can be accelerated by being subjected to the influence of an electric field so oriented as to pull the particle forward. The greater the electric potential to which the particle is subjected, the greater the acceleration, and the greater the energy gain of the particle.

A particle with a unit charge, such as an electron, which is accelerated by a field with an electric potential of one volt, gains an energy of one *electron-volt*. The electron-volt, often abbreviated to *ev*, is equal to 1.6×10^{-12} ergs. For larger units of this sort, we have the *kiloelectron-volt* (*Kev*), which is equal to 1000 electron-volts. Beyond that is the *Mev* (a million electron-volts) and the *Bev* (a billion electron-volts.)* A Bev is equal to 1.6×10^{-3} ergs. This is a small quantity of energy in ordinary terms, but it is simply enormous when we consider that it is packed into a single sub-atomic particle.

Mass can be expressed in electron-volts and subatomic masses are expressed in this way with increasing frequency. The mass of an electron is 9.1×10^{-2s} grams. This can be expressed as its equivalent in energy (as calculated by means of Einstein's mass-energy equivalence equation, $e = mc^2$, see page II–111) and turns out to be 8.2×10^{-7} ergs. This, in turn, equals 510,000 electron-volts, or 0.51 Mev.

The wavelength of electromagnetic radiation can also be expressed in electron-volts. According to the quantum theory, $e = h\nu$, where e is the energy of a quantum of electromagnetic radiation in ergs, h is Planck's constant in erg-seconds, and ν (the Greek letter "nu") is the frequency of the radiation in cycles per second.

The wavelength of that radiation (represented by λ, the Greek letter "lambda") is equal to the distance in centimeters traveled, in a vacuum, by the radiation in one second (c) divided by the number of wavelengths formed in that time—that is, by the frequency of the radiation (ν).

In other words:

$$\lambda = \frac{c}{\nu} \qquad \text{(Equation 9–2)}$$

or:

$$\nu = \frac{c}{\lambda} \qquad \text{(Equation 9–3)}$$

Substituting c/λ for ν in the quantum theory equation $e = h\nu$, we have:

$$e = \frac{hc}{\lambda} \qquad \text{(Equation 9–4)}$$

or:

$$\lambda = \frac{hc}{e} \qquad \text{(Equation 9–5)}$$

* The term "billion" has different meanings in different parts of the world. To an American, for instance, it means 1,000,000,000, but to an Englishman it means 1,000,000,000,000; and what we call a billion, they would call a "thousand million." In Great Britain, then, 1,000,000,000 electron-volts is spoken of as a "giga-electron-volt" and is abbreviated Gev.

The value of h is 6.62×10^{-27} erg-seconds, while that of c is equal to 3.00×10^{10} centimeters per second. Consequently, hc is equal to 1.99×10^{-16} ergs. We can therefore write Equation 9–5 thus:

$$\lambda = \frac{1.99 \times 10^{-16}}{e} \qquad \text{(Equation 9–6)}$$

If we substitute the value of 1.6×10^{-12} ergs (the value of one electron-volt) for e in Equation 9–6, we obtain a value of 1.24×10^{-4} centimeters. In other words, radiation with a wavelength of 1.24 microns (in the infrared range) is made up of photons with an energy of 1 ev.

It follows that one kev is the energy content of radiation with a wavelength one-thousandth as great—that is, of 1.24 millimicrons, or 12.4 angstrom units. This is in the X-ray range. Similarly, one Mev is the energy content of radiation with a wavelength of 0.0124 angstrom units, which is in the gamma-ray range.

Conversely, Equation 9–6 can be used to show that visible light has an energy content varying from 1.6 ev at the red end of the spectrum to 3.2 ev at the violet end. Ordinary chemical reactions are brought about by visible light and ultraviolet light and, in turn, produce such radiation. You can see then that ordinary chemical reactions involve energies of from not more than one to five electron-volts. It is a measure of the increased difficulty of bringing about nuclear reactions that particles with energies in the thousands of electron-volts, and even millions of electron-volts, are required for the purpose.

Particle Accelerators

Devices intended to produce subatomic particles with energies in the kev range and beyond are called *particle accelerators*. Since the energetic particles produced by these accelerators were used to disrupt atom nuclei and induce nuclear reactions, the devices were popularly called "atom-smashers," though this term has rather gone out of fashion.

The first particle accelerator to achieve useful results was one that was adapted to accelerate protons by the English physicist John Douglas Cockcroft (1897–   ) and his Irish co-worker Ernest Thomas Sinton Walton (1903–   ), in 1929.

Protons are preferable to alpha particles in that the former carry a smaller positive charge and are therefore subjected to a smaller repulsive force from atomic nuclei. In addition, protons are ionized hydrogen atoms (H^+), while alpha particles are ionized helium atoms (He^{++}); and hydrogen is both far more common and far more easily ionized than is helium.

The Cockcroft-Walton device used an arrangement of condensers to build up potentials to extraordinarily high levels (it was called a *voltage multiplier*) and to accelerate protons to energies of as high as 380 Kev.

In 1931, they were able to use such accelerated protons to bring about the disruption of a lithium nucleus:

$$_3Li^7 + _1H^1 \longrightarrow _2He^4 + _2He^4 \qquad \text{(Equation 9–7)}$$

This was the first completely artificial nuclear reaction, for here even the bombarding particles were artificially produced.

In that same year, 1931, no less than three other important types of particle accelerators were introduced.

The American physicist Robert Jemison Van de Graaf (1901–   ) built a mechanism shaped like half a dumbbell standing on end. Within it, a moving belt was so arranged as to carry positive electric charge upward and negative electric charge downward, producing a large electrostatic charge on either end. This *electrostatic generator* produced a huge potential difference that could accelerate particles to an energy of 1.5 Mev. Later such devices produced particles of still higher energies—as much as 18 Mev.

A second variety of accelerator was built up of separate tubes. This made it possible to accelerate particles by separate individual potential "kicks," instead of attempting to do it all in one powerful kick. In each tube the particle gained additional energy and took on additional velocity. Since the potential kicks were administered at equal intervals of time, the accelerating particle covered longer and longer distances between kicks, and each successive tube had to be made longer. For this reason, the *linear accelerator*, or *linac*, quickly grew inconveniently long.

The most compact arrangement for building up huge energies was the product of the American physicist Ernest Orlando Lawrence (1901–1958), who sought to save space by having the particles travel in a curved path, rather than in a straight line.

A high-temperature filament at the center of a closed flat circular vessel ionizes low-pressure hydrogen to produce protons. Opposite halves of the vessel are placed under a high potential that accelerates the protons. The poles of a magnet above and below the vessel force the protons to follow a curved path.

Ordinarily, the protons following this curved path would

eventually find themselves moving toward the positively-charged portion of the vessel and begin to slow up. However, the vessel is under an alternating potential, so that cathode and anode flip back and forth rapidly, at a carefully adjusted rate.

Each time the protons turn in such a way as to be moving toward the anode, there is a flip and the protons are moving toward the cathode after all. They are therefore pulled forward and accelerated further. (It is very like a greyhound pursuing an electric rabbit that always remains just ahead.)

As the protons accelerate, they move faster and faster, and it might be thought that they would make their turns about the vessel in less and less time. In such a case, the flip-flop of the electric field, which continues at a constant rate, would fall out of synchronization. The protons would find themselves heading toward the repulsive force of the anode, which would not be replaced by the cathode in time, and the proton would be slowed up. (This would be like the greyhound putting on a burst of speed and catching the electric rabbit.)

Fortunately, as the protons are accelerated, they naturally curve to a lesser degree under the influence of the magnetic field. They move in a larger circle and their greater velocity is just compensated for by the longer distance through which they must travel. They therefore continue to move from one half to the other in a fixed cycle that matches the alternation of the potential, spiraling outward from the center of the container as they do so. Eventually, they spiral out of a prepared exit as a stream of high-energy particles.

Linear particle accelerator

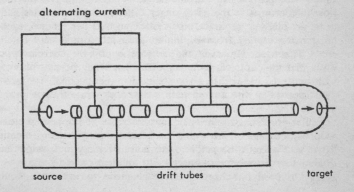

alternating current

source   drift tubes   target

Lawrence called his instrument a *cyclotron,* and even his first model, which was no more than eleven inches in diameter and intended only as a small-scale test of the principle, managed to produce particles of 80 Kev. Over the next ten years larger and larger cyclotrons were built, and particles with more than 10 Mev of energy were produced.

This perfect matching of particle movement and potential alternation works only if the mass of the particle remains unchanged. Under ordinary conditions, it does (just about), but as acceleration proceeds, particles eventually move at velocities that are sizable fractions of that of light. Acceleration begins to involve increasingly minor additions to the particle's velocity (which cannot, in any case, exceed that of light) and increasingly major additions to its mass, in accordance with the special theory of relativity (see page II–102).

As the mass of the particle increases, it takes longer than would otherwise be expected to make its semicircle, and the particle motion falls out of synchronization with the alternation of the potential. This sets a limit to the energies that can be piled on the proton, and this limit was reached by World War II.

In 1945, two men, the American physicist Edwin Mattison McMillan (1907–   ) and the Russian physicist Vladimir I. Veksler, independently suggested a way of getting around this. They showed how the alternation of the potential could be decreased gradually at just the rate required to keep it synchronized to the motion of the increasingly massive particle. The result is a *synchrocyclotron.*

A synchrocyclotron cannot produce high-energy particles continuously, for the alternation period of the potential that is suitable for particles in the late stages of acceleration is far too slow for particles in the early stages. Therefore the particles had to be produced in separate bursts of 60 to 300 per second, each burst being carried from beginning to end before a new batch could be started. However, the increase in possible energies was well worth the cut in total quantity. The first synchrocyclotron was built in 1946, and within a few years instruments capable of producing particles with energies up to 800 Mev made their appearance.

The problem of relativistic mass increase had appeared even sooner in connection with electron acceleration. Electrons are so light that they must be made to move at extremely high velocities to achieve even moderate energies. If an electron is to attain an energy of even one Mev, it must be made to move at about

270,000 kilometers per second, which is better than 9/10 the speed of light. At that speed, the mass of the electron is 2.5 times what it would be if the particle were at rest. Using the cyclotron principle on the electron is therefore impractical, for the electron would lose synchronization at very low energies.

A solution to this was found even before the principle of the synchrocyclotron was established. In 1940, the American physicist Donald William Kerst devised an accelerator in which the electrons were made to move in a circle through a doughnut-shaped vessel. As they gained velocity, the strength of the magnetic field that made the electrons move circularly was increased. Since the increase in magnetic field intensity (which tended to make the electrons move in a more sharply curved path) was matched with the increase in the electrons' mass (which tended to make them move in a less sharply curved path), the net result was to keep the electrons moving in the same path. At a given moment, a sudden change in the magnetic field hurled a burst of high-energy electrons out of the instrument. Because beta particles are a natural example of high-energy electrons, Kerst called his instrument the *betatron*. Kerst's first instrument produced electrons with an energy of 2.5 Mev, and the largest betatron built since produces electrons with an energy of 340 Mev.

Electrons whirling very rapidly in circular orbits are sharply accelerated toward the center and, as Maxwell's theory of electromagnetic phenomena would require, give off energy in the form of radiation. This sets a limit to how much energy can be pumped into electrons by any device requiring the particles to travel in circles. (This phenomenon is less marked for protons, which for a given energy need not travel so quickly nor be subjected to so great a consequent acceleration.)

For this reason new efforts are now being made to manufacture linear accelerators long enough (and a two-mile-long unit is being planned) to produce electrons with energies up to 20,000 Mev, or 20 Bev.

The synchrocyclotron has one defect, not in theory, but in practice. As the particle spirals outward, it sweeps through curves of greater and greater radius, and the magnet must be large enough to cover the maximum radius. Magnets of the proper enormous size were a bottleneck in construction of larger machines.

There was an advantage then in adjusting the magnetic field to allow protons to travel in circles rather than in spirals. The design was such that "strong-focusing" was introduced, making the proton stream hold together tightly in as narrow a beam as possible. In this way, *proton synchrotrons* and *electron synchrotrons* were built.

By 1952, proton synchrotrons capable of producing particles in the Bev range were built. There is such a device at the University of California; it is appropriately called the *Bevatron* and can produce protons with energies of 6.2 Bev.

In the 1960's, particularly large strong-focusing accelerators were built (one in Geneva, and one at Brookhaven, Long Island) and are capable of producing protons with energies in excess of 30 Bev. Still larger machines are envisaged, but the plans are, of necessity, colossal. Present large accelerators are three city blocks in diameter.

CHAPTER 10

Artificial Radioactivity

Radioisotopes

The first nuclei produced by artificial transmutation were stable ones that exist in the elements as found in nature. Examples are the oxygen-17 produced by Rutherford and the helium-4 produced by Cockcroft and Walton.

This precedent was shattered in 1934 through the work of the French physicist Frédéric Joliot-Curie (1900–1958) and his wife, Irène (1897–1956)—who were the son-in-law and daughter of Pierre and Marie Curie, the discoverers of radium.

The Joliot-Curies continued Rutherford's work on alpha particle bombardment of nuclei. In bombarding aluminum, they found emissions of protons and neutrons, emissions which ceased when the alpha particle bombardment was interrupted. Another type of radiation* did not cease but fell off in an exponential manner, with a half-life of 2.6 minutes. It seemed quite plain that something in the aluminum that had not originally been radioactive had become radioactive as a result of the bombardment.

The following equation describes what happens when aluminum-27 absorbs an alpha particle and emits a proton:

* The nature of this radiation will be discussed later in the book; see page 224.

$$_{13}Al^{27} + _2He^4 \longrightarrow _{14}Si^{30} + _1H^1 \qquad \text{(Equation 10-1)}$$

or:

$$Al^{27}(\alpha,p)Si^{30}$$

Silicon-30 is a stable isotope, occurring in silicon with a relative abundance of just about 3 percent.

But under the bombardment aluminum also emits neutrons. It follows then that a reaction might be taking place in which the target aluminum nucleus absorbs an alpha particle and emits a neutron, making a net gain of two protons and one neutron. In such an (α,n) reaction, then, the atomic number is increased by two rather than by one, and aluminum is converted to phosphorus rather than to silicon. The equation can be written:

$$_{13}Al^{27} + _2He^4 \longrightarrow _{15}P^{30} + _0n^1 \qquad \text{(Equation 10-2)}$$

or:

$$Al^{27}(\alpha,n)P^{30}$$

But phosphorus as it occurs in nature is made up of a single isotope, phosphorus-31. No other stable phosphorus isotope is known, and it is to be presumed that if any other phosphorus isotope is synthesized in the course of a nuclear reaction, it would be radioactive; and it is because of this radioactivity (combined with a short half-life) that it does not occur in nature.

The Joliot-Curies confirmed the presence of radioactive phosphorus in the aluminum by dissolving the metal and allowing it to undergo reactions that would put any phosphorus present into the form of either a gaseous compound or a solid precipitate. Sure enough, the radioactivity was found in the gas or the precipitate.

Phosphorus-30 was the first isotope to be produced in the laboratory that did not occur on earth naturally, and it is also the first example of *artificial radioactivity*.

It was by no means the last. Over the next generation, nuclear reactions induced in the laboratory produced over a thousand such artificial isotopes. Since every single one of those so produced is radioactive, they are often called *radioisotopes*.

Radioisotopes of every stable element have been formed, sometimes in considerable number. In the case of cesium, for instance, which has a single stable isotope, cesium-133, no less than twenty different radioisotopes have been formed, with mass numbers of from 123 to 148.

None of the radioisotopes so produced have half-lives long

enough to allow them to remain in the earth's crust over the planet's lifetime. Some of the half-lives are long by human standards to be sure (cesium-135 has a half-life of 2,000,000 years), but none are long enough.

One might suspect that at the time the matter making up the solar system was created, all conceivable nuclear arrangements were brought into existence. Those that happened to be stable, and those that were only very slightly radioactive (as, for instance, potassium-40 and uranium-238), survived. And indeed it seems quite likely that all stable or nearly stable isotopes that can exist do exist on earth, and the chances are virtually zero that an unknown stable or nearly stable isotope will ever be discovered.

As for those isotopes that are sufficiently unstable to have half-lives of less than 500,000,000 years, they may have also been formed, but they broke down and disappeared, some rapidly and some less rapidly. It is only the labor of the physicist that now brings them back to life.

The Biochemical Uses of Isotopes

Once physicists began isolating rare isotopes and synthesizing new ones, it became possible to prepare chemical compounds containing them. If the isotopes could be prepared cheaply enough, then the compounds containing them could be used in chemical experiments in quantity.

The first isotope to be used in comparatively large-scale experimentation in this manner was the stable hydrogen-2, which could be prepared in the form of "heavy water" (see page 144).

By carrying out organic chemical reactions in heavy water, it was possible to prepare other compounds with molecules containing one or more atoms of hydrogen-2. If such compounds were allowed to take part in chemical reactions, their ultimate fate could be determined by isolating the products and checking to see which of them contained hydrogen-2. A compound containing a more-than-normal amount of a rare isotope may therefore be said to be a *tagged compound*, and the abnormal atom itself, an *isotopic tracer*.

This technique is particularly important where the tagged compound is one that ordinarily undergoes chemical changes in living tissue, for then it can be followed through the rapid and extraordinarily complicated transformations that take place there. Beginning in 1935, the German-American biochemist Rudolf Schoenheimer (1898–1941) carried on such experiments, making use of fat molecules rich in hydrogen-2. This introduced a veritable revolution in biochemistry, for it quickly became possible to work out details of tissue reactions that might otherwise have remained impenetrable.

Schoenheimer, and others as well, also worked with the heavier isotopes of nitrogen and oxygen. These were nitrogen-15 and oxygen-18, with relative abundances of 0.37 percent and 0.20 percent, respectively. Both are rare enough in nature to serve as effective tracers when used in concentrated fashion.

The production of radioisotopes made possible an even greater sensitivity in the use of isotopic tracers, for radioactive isotopes can generally be detected more easily, more quickly, and in much smaller concentration than can stable isotopes.

Radioactive tracers were used as early as 1913 by the Hungarian physicist Georg von Hevesy (1885–   ). At the time, the only radioactive isotopes that were available were those that were members of the various radioactive series. Hevesy made use of lead-210 in determining the solubility of very slightly soluble lead compounds. (He could determine the fraction of lead-210 that went into solution by measuring the radioactivity of the solution before and after, and it seemed reasonable to assume that this fraction held good for all lead isotopes generally.)

In 1923, Hevesy tagged a lead compound with lead-212 and studied the uptake of lead by plants. This was the first biological application of isotopic tracers. However, lead is not a compound that occurs naturally in living tissue; indeed, lead is an acute poison. The behavior of tissue in the presence of lead is not necessarily normal. The use of radioisotopes of the more biologically useful elements did not become really large-scale until after World War II, when methods for preparing such isotopes in quantity were developed.

One unavoidable shortcoming of the radioisotope technique is that few good radioisotopes are available for those elements most common in tissue. The four elements making up over 90 percent of the soft tissues of the body are carbon, hydrogen, oxygen and nitrogen. In the case of nitrogen, the most long-lived radioisotope known is nitrogen-13, which has a half-life of ten minutes. That means that once nitrogen-13 is formed, it must be incorporated into a suitable compound, made available to the tissues, meet whatever fate it will, and have its products isolated and investigated—all in a matter of half an hour or so. Even after a mere half an hour, the radioactivity is already only 1/8 what it was to start with.

For oxygen the situation is much worse because the most long-lived radioisotope known here is oxygen-15, which has a half-life of only two minutes.

In the case of carbon, the most long-lived radioisotope known before 1940 was carbon-11, which has a half-life of twenty minutes. This was a borderline situation. It left little time for maneuver, but of all the elements in living tissue, carbon was by far the most important; biochemists therefore worked out methods for squeezing information out of experiments using compounds tagged with carbon-11, despite the tight time-limit enforced by the short half-life.

It was not expected that any longer-lived carbon isotope would be discovered. In 1940, however, a new radioisotope of carbon was discovered as the result of the bombardment of carbon itself with *deuterons* (the nuclei of deuterium, H^2.)

A deuteron is made up of a proton and a neutron, and the carbon atoms undergoing the deuteron bombardment give off protons, retaining the neutrons. In a (d,p) reaction, the atomic number remains unchanged, but the mass number increases by one. Carbon is made up of two stable isotopes, carbon-12 and carbon-13. The former is converted to the latter by a (d,p) reaction, but the latter undergoes the following:

$$_6C^{13} + {}_1H^2 \longrightarrow {}_6C^{14} + {}_1H^1 \qquad \text{(Equation 10–3)}$$

or:

$$C^{13}(d,p)C^{14}$$

Carbon-14 is radioactive and has the unexpectedly long half-life of 5770 years. In terms of the duration of any laboratory experiment likely to be conducted with carbon-14, its radioactivity rate can be considered constant. Numerous biological and biochemical experiments were conducted with compounds tagged with carbon-14, and it is undoubtedly the most useful single radioisotope.

In 1946, the American chemist Willard Frank Libby (1908–) pointed out that carbon-14 should exist in nature as a result of nuclear reactions indirectly induced in the nitrogen-14 present in the atmosphere by energetic radiations from outer space.* This reaction is, in essence, the gaining of a neutron and the loss of a proton. In such an (n,p) reaction, there is no net change in the mass number, but a decrease of one in the atomic number. Thus:

---

* These radiations, called cosmic rays, will be taken up on page 217.

$$_7N^{14} + _0n^1 \longrightarrow {}_0C^{14} + _1H^1 \qquad \text{(Equation 10-4)}$$

or:

$$N^{14}(n,p)C^{14}$$

Carbon-14 is continually being formed in this way, and it is also continually breaking down after being formed. There is an equilibrium between the two processes, and the carbon-14 in the atmosphere (occurring in part of its carbon dioxide content) is at a constant, though very low, level.

Libby further pointed out that since plant life constantly absorbed and made use of carbon dioxide, its tissues ought to contain a constant, though very low, concentration of carbon-14, and so ought animal tissues, since animals feed on plants (or on other animals which feed on plants).

The constant concentration of carbon-14 in tissue was only maintained, however, while that tissue was alive, since only then was radioactive carbon being continually incorporated, either by absorption of atmospheric carbon dioxide or by the ingestion of food. Once a creature dies, intake of carbon-14 ceases, and the amount already present begins to decrease in a fixed manner.

Anything that was once part of a living organism can be analyzed for its carbon-14 content, and the time-lapse since life ended can be determined. This method of *radiocarbon dating* has been much used in archaeology. Wood from an old Egyptian tomb was found to be roughly 4800 years old, for instance, while wood from an old Etruscan tomb was about 2730 years old. The age of the Dead Sea Scrolls has been confirmed in this manner.

Wood from ancient trees knocked over by advancing glaciers can be tested, as can driftwood that once lay on the shores of lakes formed from melting glaciers. Scientists were surprised to discover that the last advance of the ice sheets that covered much of North America began but 25,000 years ago and reached its maximum extent about 18,000 years ago. This was not as long ago in the past as had previously been thought. Even as recently as 10,000 years ago, the retreating glaciers made a new partial advance, and it wasn't until 6000 B.C. (when men were already preparing to build their first civilizations) that the glaciers finally disappeared from the Great Lakes regions.

The (d,p) reaction that had led to the discovery of carbon-14 had earlier led to the discovery of the only radioisotope of hydrogen. In 1934, the Australian physicist Marcus Laurence Elwin Oliphant (1901–   ) had bombarded deuterium gas with deuterons. The heavy hydrogen nucleus (H^2) was thus both target and bombarding particle:

$$_1H^2 + {}_1H^2 \longrightarrow {}_1H^3 + {}_1H^1 \qquad \text{(Equation 10–5)}$$

or:

$$H^2(d,p)H^3$$

The hydrogen-3 formed in this way has the unexpectedly long half-life of 12.26 years. It has been named *tritium* (from a Greek word meaning "third") and its nucleus, composed of one proton and two neutrons, is a *triton*. Tritium is also formed naturally in the atmosphere through the action of high-energy radiation, so extremely small quantities are present in ordinary water. In very special cases, the decline in tritium content can be used in dating.

Units of Radioactivity

In using radioisotopes, what counts is not the mass alone but also the breakdown rate, for it is the latter that governs the quantity of particles being emitted per unit mass, and it is those particles which must be detected.

The breakdown rate (R_b) of a radioisotope can be expressed as follows:

$$R_b = \frac{0.693N}{T} \qquad \text{(Equation 10–6)}$$

where N is the total number of radioactive atoms present, and T is the half-life in seconds.

Let's consider a gram of radium. The mass number of the most long-lived radium isotope (and the one almost invariably meant when the unqualified word "radium" is used) is 226. This means that 226 grams of radium contains Avogadro's number of atoms, 6.023×10^{23} (see page 20). One gram of radium therefore contains Avogadro's number divided by 226, or 2.66×10^{21} atoms. The half-life of radium-226 is 1620 years, or 5.11×10^{10} seconds.

Substituting 2.66×10^{21} for N in Equation 10–6, and 5.11×10^{10} for T, we find a value of 3.6×10^{10} for R_b. This means that in a gram of radium, 36,000,000,000 atoms are breaking down each second.

In 1910, it was decided that the number of atomic breakdowns in one gram of radium be taken as a unit called a *curie*, in honor of the discoverers of radium. At the time, the calculation of this figure yielded the value of 37,000,000,000 breakdowns per second. One therefore defines 1 curie as equal to 3.7×10^{10} atomic breakdowns per second. The number of breakdowns per gram of radioisotope is its *specific activity*. Thus the specific activity of radium is 1 curie per gram.

What about other isotopes? The breakdown rate is inversely proportional to the half-life. The longer the half-life, the fewer the atomic breakdowns per second in a given quantity of radioisotope, and vice versa. The breakdown rate is therefore proportional to T_r/T_1 where T_r is the half-life of radium-226 and T_1 is the half-life of the other isotope.

For a fixed breakdown rate, the actual number of breakdowns in a gram of isotope is inversely proportional to the mass number of the isotope. If the isotope is more massive than radium-226, fewer atoms will be squeezed into one gram, and there will be fewer breakdowns in that one gram. The reverse is also true. The number of breakdowns will be proportional to M_r/M_1, where M_r is the mass number of radium-226 and M_1 is the mass number of the isotope.

The specific activity (S_a) of a radioisotope—that is the number of breakdowns per second in one gram, as compared with that in one gram or radium—depends on the half-lives and mass numbers as follows:

$$S_a = \frac{T_r M_r}{T_1 M_1}$$

(Equation 10-7)

Since the half-life of radium-226 is 5.11×10^{10} seconds and its mass number is 226, the numerator of Equation 10-7 is equal to $226(5.11 \times 10^{10})$, or 1.15×10^{13}. Therefore:

$$S_a - \frac{1.15 \times 10^{13}}{T_1 M_1}.$$

(Equation 10-8)

Thus, carbon-14, which has a half-life of 5770 years, or 1.82×10^{11} seconds, and a mass number of 14, has 2.55×10^{12} for its value of $T_1 M_1$. If 1.15×10^{13} is divided by 2.55×10^{12}, we find that the specific activity of carbon-14 is 4.50 curies per gram. Carbon-14 has a longer half-life than radium-226, and that cuts down its breakdown rate. However, carbon-14 is a much lighter atom than radium-226; consequently, there are many more of the former per gram, and the actual number of breakdowns in that gram is greater than in the case of radium-226, despite the lower breakdown rate.

On the whole, most radioisotopes used in the laboratory have half-lives shorter and mass numbers smaller than that of radium, so that the specific activity is generally very high. Thus, carbon-11 has a half-life of 20.5 minutes, or 1230 seconds, a mass number of 11, and a specific activity of 850,000,000 curies per gram.

To be sure, gram lots of these isotopes are not used. They are generally not available in such quantities in the first place and would be highly dangerous if they were. Besides, such quantities are not needed. Particle detection is so delicate that the curie turns out to be a unit too large for convenience, and one more often speaks of *millicuries* (1/1000 of a curie) or *microcuries* (1/1,000,000 of a curie). Thus a microgram of carbon-11 is equivalent to 850 microcuries.

Even a microcurie represents a breakdown rate of 36,000 per second. Under the best conditions, four breakdowns per second may be detected with reasonable precision. This would represent 1/9000 of a microcurie, or 1.1×10^{-10} curie.

The use of the curie is made inconvenient to some extent by the fact that it represents a large and "uneven" number of atomic breakdowns per second. It has been suggested that the *rutherford* be used instead (named in honor of the discoverer of the nuclear atom). One rutherford is defined as a million atomic breakdowns per second.

This means that 1 curie = 37,000 rutherfords and that 1 rutherford = 270 microcuries.

Neutron Bombardment

As soon as the neutron was discovered, it occurred to physicists to use it as a bombarding particle to bring about nuclear reactions (and it was this, really, which eventually led to the wholesale production of radioisotopes). However, an apparent disadvantage of the neutron in such a role is its lack of charge. This means it cannot be accelerated by the electric fields used by all particle accelerators.

One way out of this dilemma was provided in 1935 by the American physicist John Robert Oppenheimer (1904–   ), who suggested the use of a deuteron instead. The deuteron is made up of a proton and a neutron in comparatively loose combination. A deuteron, with a charge of +1, can be accelerated. As the energetic deuteron approaches the positively-charged target nucleus, the proton component is repelled, sometimes strongly enough to break the combination. The proton veers off, but the neutron, unaffected by the repulsion, continues on, and if its aim is true, may be absorbed by the nucleus. The result resembles a

(d,p) reaction of the type shown in Equations 10–3 and 10–5.

However, the inability to accelerate neutrons themselves is by no means a fatal defect. Indeed, it scarcely matters. A neutron, being neither attracted nor repelled by electric charge, can strike a nucleus (if aimed in the correct direction) regardless of how little energy it carries.

During the 1930's, streams of neutrons were produced from atoms subjected to bombardment by alpha particles. An alpha particle source mixed with beryllium served as a particularly useful neutron source.

A neutron may be absorbed by a target nucleus without the immediate emission of some other particle. Instead, the nucleus reaches an excited state as a result of absorbing the kinetic energy of the neutron and simply radiates off that excess energy as a gamma-ray photon. This is a (n,γ) reaction. The energy may not be written explicitly into the equation representing this reaction, thus:

$$_{48}Cd^{114} + {_0}n^1 \rightarrow {_{48}}Cd^{115}$$
(Equation 10–9)

or:

$$Cd^{114} (n,\gamma) Cd^{115}$$

The neutron, even more surely than the deuteron, can thus be used to produce higher isotopes of a target element.

It often happens that the higher isotope so produced is radioactive and breaks down by emitting a beta particle. This does not affect the mass number but raises the atomic number by one. Cadmium-115, for instance, is a beta-emitter with a half-life of 43 days, and is converted to indium-115.

If cadmium-116 had been bombarded with neutrons and converted to cadmium-117, there would follow a double change. Cadmium-117 is a beta-emitter with a half-life of about three hours and becomes indium-117, which is a beta-emitter with a half-life of about two hours and is converted, by beta particle emission, to the stable tin-117.

In many cases, then, neutron bombardment can produce an element one or two atomic numbers higher than the target element. The efficiency with which this may be achieved depends upon the probability of a (n,γ) reaction taking place. This probability can be dealt with as follows:

Imagine a target material one square centimeter in area and containing N atomic nuclei. Suppose it is bombarded by I particles per particle and that A atomic nuclei are hit per second.

The part of the target actually struck by the particles in one second is therefore A/N.

That part, however, is hit by all I particles lumped together. The part of the target hit by a single particle has to be A/N divided by I. The size of the target hit by a single particle is the *nuclear cross section,* which is symbolized as σ (the Greek letter "sigma"). We can say then that:

$$\sigma = \frac{A}{NI} \qquad \text{(Equation 10-10)}$$

By this analysis it would seem that in order to induce re-action a single bombarding particle must strike a particular area, σ square centimeters in size, centered about a particular target nucleus. The value of the nuclear cross section, as worked out by Equation 10-10, usually comes out in the neighborhood of 10^{-24} square centimeters. For convenience, nuclear physicists have defined 1 *barn* as equal to 10^{-24} square centimeters. (The story is that the name of the unit arose out of a statement that on the subatomic scale hitting an area 10^{-24} square centimeters in size, was like hitting the side of a barn on a familiar everyday scale.)

The value of the nuclear cross section varies with the nature of the target nucleus and with the nature of the bombarding particle. The Italian physicist Enrico Fermi (1901–1954) found in 1935 that neutrons became more efficient in bringing about nuclear reactions after they had been passed through water or paraffin. The nuclear cross section for bombarding neutrons on a given target nucleus increased, in other words, after the neutron's passage through water or paraffin.

In passing through water or paraffin, neutrons collided with light atoms that were particularly stable and therefore had little tendency to absorb an additional neutron. (They had low cross sections for neutron absorption, in other words.) As a result, the neutron bounced off.

When two objects bounce, there is usually a redistribution of kinetic energy between them. If one of the objects is moving and one is at rest, the moving object loses some energy and the object at rest gains some. The division of energy is most likely to be equal if the two bouncing objects are more or less equal in mass.

We can see this on a large scale if we imagine ordinary objects in place of subatomic particles. If a moving billiard ball collides with a ping-pong ball (the case of a neutron striking an electron), the ping-pong ball will bounce away vigorously, but the billiard ball loses little energy and goes on its way as before. On the other hand, if a moving billiard ball collides with a cannonball (the case of a neutron striking the nucleus of a lead atom), the billiard ball merely bounces, retaining its energy, while the cannonball is virtually unaffected. However, if a billiard ball strikes another billiard ball, it is quite likely that the two will end with roughly equal energies.

Consequently, a neutron is most efficiently slowed if it bounces off light nuclei such as those of hydrogen, beryllium or carbon, and it does this when passing through compounds, such as water and paraffin, made up of these light atoms. Such substances act as *moderators*. In the end, neutrons can be slowed until they are moving at no more than the velocity of atmospheric atoms and molecules under the influence of the local temperature (see page I-205). These are *thermal neutrons*.

Why then should nuclear cross sections rise as neutrons are slowed down? To answer this, we must remember that while neutrons have some particle-like properties, they also have wave-like properties. It had already been shown in the 1920's that electrons exhibit the wave-like properties predicted for them by de Broglie (see page 102), but there remained some question as to whether this might not apply only to charged particles. In 1936, it was shown that neutrons passing through crystals were diffracted, and the wave-particle duality was demonstrated for all matter and not for electrically charged matter only.

As a particle slows down, it loses energy. In its wave aspect this lowering of energy is represented by an increase in wavelength. A neutron therefore "spreads out" and grows "fuzzier" as it slows down. The larger, slow neutron is more likely to strike a nucleus than the smaller, fast neutron and is therefore more likely to bring about a nuclear reaction. It is also true that a slow neutron remains in the vicinity of the target nucleus a longer interval of time than a fast neutron, and this, too, encourages reactions.

Synthetic Elements

The development in the 1930's of new methods for bringing about nuclear reactions led to the formation not only of isotopes not found in nature, but of elements not found there.

In the 1930's, there existed just four gaps in the list of elements from atomic numbers 1 to 92 inclusive. These were the elements of atomic numbers 43, 61, 85 and 87.

The first of the four gaps to be filled was that of element number 43. Lawrence, the inventor of the cyclotron, had exposed molybdenum to a stream of accelerated deuterons, and it was possible that element number 43 had been produced in a (d,n) reaction:

$$_{42}Mo^{98} + _1H^2 \longrightarrow _{43}X^{99} + _0n^1 \qquad \text{(Equation 10–11)}$$

or:

$$Mo^{98}(d,n)X^{99}$$

A sample of the irradiated molybdenum reached the Italian physicist Emilio Segrè (1905–   ) in 1937. He tested it by chemical methods to see if any part of the new radioactivity would follow the course to be expected of element number 43. It did; and it was amply confirmed that element 43 existed in the molybdenum. Since it was the first element to be discovered as the result of man-made nuclear reactions, it was named *technetium* ("artificial").

Not only was technetium the first man-made element, but it was also the first case of a light element (one with an atomic number of less than 84) that lacked any stable isotope whatever. There are no less than three technetium isotopes with quite long half-lives—technetium-97, 2,600,000 years; technetium-98, 1,500,000 years; and technetium-99, 210,000 years. However, no isotopes are completely stable. Since none of the half-lives of these isotopes are long enough to survive the ages of the earth's existence and since no technetium isotopes are part of a radioactive series, there are no measurable quantities of technetium present in the earth's crust.

In 1939, the French chemist Mlle. M. Perey discovered an isotope of element 87 among the breakdown products of uranium-235. She named it *francium*, after her native land. Element 85 was also later detected in the radioactive series, but as early as 1940 it had been produced artificially by the bombardment of bismuth with alpha particles and was named *astatine* ("unstable"). The reaction was:

$$_{83}Bi^{209} + _2He^4 \longrightarrow _{85}At^{211} + _0n^1 + _0n_1$$
$$\text{(Equation 10–12)}$$

or:

$$Bi^{209}(a,2n)At^{211}$$

(Segrè had by now come to the United States and was a member of the group that isolated astatine.)

Element number 61 was discovered in 1948 (under circumstances to be described later in the book) by a team working under the American chemist Charles DuBois Coryell (1912–) and was named *promethium*. It was the second case of a light element without stable isotopes. Indeed, its most long-lived isotope, promethium-145, has a half-life of only 18 years.

By 1948, therefore, the periodic table had finally been filled up and its last gap removed. Meanwhile, however, the table had opened at its upper end. Fermi, mindful of the ability of the neutron to raise the atomic number of a target nucleus by one or two, had since 1934 been bombarding uranium with neutrons.

He felt that uranium-239 might be formed from uranium-238. By emitting beta particles this might become an isotope of element 93 and then possibly of element 94. He thought at first that he had actually demonstrated this and referred to the hypothetical element 93 as "Uranium X."

The discovery of uranium fission (see page 192) showed that Fermi had done far more then prepare element 93 and, for a while, element 93 was forgotten. However, when the furor of fisson had died down a bit, the question of element 93 was taken up again. The formation of uranium-239 was not the chief result of the neutron bombardment of uranium, nor the most important, but it was a result. It took place.

This was finally demonstrated in 1940 by the American physicist Edwin Mattison McMillan and his colleague, the American chemist Philip Hauge Abelson (1913–). They traced radioactivity showing a 2.3 day half-life and found it belonged to an isotope with an atomic number of 93 and a mass number of 239. Since uranium had originally been named for the planet Uranus, the new element beyond uranium was named *neptunium*, for Neptune, the planet beyond Uranus.

It seemed quite likely that this isotope, neptunium-239, was a beta particle emitter and decayed to an isotope of element number 94. However, the isotope so produced was apparently so weakly radioactive as to be difficult to detect in small quantities. By the end of the year, however, McMillan and a new assistant, the American chemist Glenn Theodore Seaborg (1912–), bombarded uranium with deuterons and formed neptunium-238:

$$_{92}U^{238} + {_1}H^2 \longrightarrow {_{93}}Np^{238} + {_0}n^1 + {_0}n^1 \quad \text{(Equation 10–13)}$$

or:

$$U^{238}(d,2n)Np^{238}$$

Neptunium-238 emitted a beta particle and formed an isotope of element 94, one which was indeed radioactive enough to detect. The new element was named *plutonium*, after the planet Pluto, which is beyond Neptune.

Once plutonium was formed in sufficient quantity, it was bombarded with alpha particles, and in 1944 a research team headed by Seaborg formed isotopes of element 95 (*americium*, for America) and 96 (*curium*, for the Curies).

Still higher elements were formed by Seaborg's group. In 1949 and 1950, elements 97 and 98 were formed by the bombardment of americium and curium with alpha particles. Element 97 was named *berkelium* and element 98 was named *californium*, after Berkeley, California, where the work was done.

Elements 99 and 100 were formed in the laboratory in 1954, but two years earlier, in 1952, isotopes of these elements were found in the residue of a hydrogen bomb test explosion (see page 208) at a Pacific atoll. By the time these discoveries were confirmed and the announcement was made, both Einstein and Fermi had died, and in their honor, element 99 was named *einsteinium* and element 100, *fermium*.

In 1955, element 101 was formed by the bombardment of einsteinium with alpha particles, and was named *mendelevium*, in honor of Mendeleev, the discoverer of the periodic table. In 1957, the discovery of element 102 was announced at the Nobel

TABLE XI—*Transuranium Elements*

Atomic Number	Element	Mass Number of Most Long-Lived Isotope	Half-Life	
93	Neptunium	[237]	2,140,000	years
94	Plutonium	[242]	37,900	years
95	Americium	[243]	7,650	years
96	Curium	[247]	c. 40,000,000	years
97	Berkelium	[247]	c. 10,000	years
98	Californium	[251]	c. 800	years
99	Einsteinium	[254]	480	days
100	Fermium	[253]	c. 4.5	days
101	Mendelevium	[256]	1.5	hours
102	Nobelium	[253]	c. 10	minutes
103	Lawrencium	[257]	8	seconds

Institute in Stockholm and was named *nobelium*,* and in 1961, element 103 was identified and named *lawrencium* in honor of the discoverer of the cyclotron, who had died some years earlier. In 1964, Soviet physicists announced the formation of element 104, but this has not yet been confirmed.

The elements beyond uranium are generally spoken of as the *transuranium elements*. Nearly a hundred isotopes of these elements have been formed. In Table XI, the most long-lived known isotopes of these elements are presented.

The chief theoretical interest in these elements is in the light they have thrown on the higher reaches of the periodic table. Before the discovery of the transuranium elements, thorium had been placed under hafnium in the periodic table; protactinium under tantalum; and uranium under tungsten. There was some chemical evidence in favor of this arrangement.

Working on this basis, when neptunium was discovered it should have fitted under rhenium. However, the chemical properties of neptunium revealed themselves almost at once to be much like uranium, and the other transuranium elements agreed in this respect. It turned out (as Seaborg was first to suggest) that the elements from actinium on formed a new "rare earth" series (see page 18) and should be fitted under the first series from lanthanum on. This is done in the periodic table presented on page 16.

The first series, from lanthanum to lutetium inclusive, is now called the *lanthanides* after the first member. Analogously, the second series, from actinium to lawrencium inclusive, is that of the *actinides*. Lawrencium is the last of the actinides and chemists are quite certain that when element 104 is obtained in quantity sufficient for its chemical properties to be studied it will turn out to resemble hafnium.

While a few of the transuranium isotopes have half-lives that are long in human terms, none are long in geologic terms, and none have survived over the eons of earth's history. (Nevertheless, traces of neptunium and plutonium have been located in uranium ores. They have arisen from the reaction of neutrons— occurring naturally in air as a result of the nuclear reaction induced by high-energy radiation from outside earth—with uranium.)

Neptunium-237 is of particular interest. Its mass number

* Attempts to duplicate the Swedish work failed. Element 102 has been formed by other methods than those described at the Nobel Institute, and the name "nobelium" is not as yet officially accepted.

divided by 4 leaves a remainder of 1, so that it belongs to the $4n + 1$ group of mass numbers, the group for which there is no naturally-occurring radioactive series (see page 132). With a half-life of over two million years, it is the longest-lived member (as far as is known) of this group. It can serve therefore as parent atom for a *neptunium* series. It gives rise to a series of daughter atoms that do not duplicate any of the products of the other three series, see Table XII.

The most distinctive feature of the neptunium series is that it ends with bismuth rather than with lead, as do the other three series. Naturally, since the parent atom of the series could not survive through earth's history, neither could any of the shorter-lived daughter atoms. The entire series is extinct except for the final stable product, bismuth-209.

TABLE XII— *The Neptunium Series*

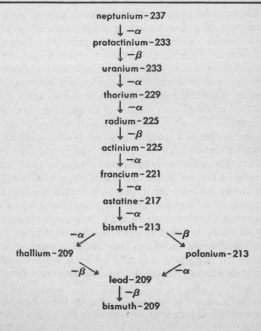

CHAPTER 11

Nuclear Structure

Nucleons, Even and Odd

With the entire list of isotopes, stable and unstable, spread before us, it is possible to make certain statements about nuclear structure.

To begin with, one can have an atom with a single proton as its nucleus; that is the case with hydrogen-1. No nucleus, however, can contain more than one proton without also containing a neutron. Among the elements with small atoms, stable nuclei tend to be made up of equal or nearly equal numbers of protons and neutrons. Thus, hydrogen-2 contains one of each; helium-4, two of each; carbon-12, six of each; oxygen-16, eight of each: sulfur-32, sixteen of each; and calcium-40, twenty of each.

The trend does not persist. All stable nuclei more massive than calcium-40 contain more neutrons than protons, and the unbalance becomes more marked as the mass number increases. Thus, the most common iron isotope, iron-56, contains 26 protons and 30 neutrons, for a neutron/proton (n/p) ratio of 1.15. The most common silver isotope, silver-107, contains 47 protons and 60 neutrons for an n/p ratio of 1.27. The only stable bismuth isotope, bismuth-209, which has the distinction of being the most massive of the stable isotopes, contains 83 protons and 126 neutrons, for an n/p ratio of 1.52; the most massive naturally-occurring isotope, uranium-238, with 92 protons and 146 neutrons, has an n/p ratio of 1.59.

Apparently, as more and more protons are packed into the nucleus, a larger and larger excess of neutrons is required to keep the nucleus stable. By the time 84 protons, or more, exist in the nucleus, no number of neutrons will suffice for stability. (And, of course, too many neutrons are as bad as too few.)

It seems quite clear that the existence of protons in pairs has a stabilizing effect on the nucleus. Of the nuclei containing more than one nucleon, those with protons in pairs (and therefore possessing an even atomic number) are the more widespread in the universe. Thus, six elements make up about 98 percent of the planet we live on (counting its interior as well as its crust) and these are: iron, oxygen, magnesium, silicon, sulfur and nickel. The atomic numbers are 26, 8, 12, 14, 16, and 28 respectively —all even.

This is reflected also in the ease with which even numbers of protons are stabilized as compared with odd numbers. For elements with atomic numbers over 83, no number of neutrons will suffice to stabilize the nucleus, but for two elements in this group, stability is nearly achieved. They are thorium and uranium, with atomic numbers of 90 and 92, both even. On the other hand, there are only two elements with atomic numbers under 83 that possess no stable isotopes. These are technetium and promethium, with atomic numbers 43 and 61, both odd.

Consider next the number of isotopes per element. There are 21 elements possessing only one naturally-occurring isotope. Of these, two have even atomic numbers: beryllium (atomic number 4) and thorium (atomic number 90). The other 19 all have odd atomic numbers. Then there are 23 elements with only two naturally-occurring isotopes. Again two of these have even atomic numbers: helium (atomic number 2) and uranium (atomic number 92). And again the other 21 all have odd atomic numbers.

Indeed, it would appear that the possession of an odd number of protons in the nucleus makes stabilization so touch-and-go that only one particular number of neutrons, or at most two, will do. Only a single element of odd atomic number possesses more than two naturally-occurring isotopes, and this is potassium (atomic number 19). It has three isotopes: potassium-39, potassium-40, and potassium-41. Of these three, however, potassium-40 is slightly radioactive and quite rare.

On the other hand, all but four of the naturally-occurring elements with even atomic numbers possess more than two naturally-occurring isotopes; indeed, tin (atomic number, 50) possesses ten. It is as though the possession of an even number of protons makes stabilization so easy that it is possible to carry it through with any of a wide variety of neutron numbers.

Neutrons also seem to occur most readily in pairs. Of the six elements earlier referred to as making up 98 percent of the earth, the most common isotopes are iron-56, oxygen-16, magnesium-24, silicon-28, sulfur-32, and nickel-58. The proton-neutron contents are 26-30, 8-8, 12-12, 14-14, 16-16, and 28-30. In every case there are even numbers of both protons and neutrons ("even-even nuclei").

Among the elements of odd atomic number which possess only one naturally-occurring isotope, in every case the single isotope possesses an even number of neutrons ("odd-even nuclei"). Examples are fluorine-19 (9 protons, 10 neutrons), sodium-23 (11 protons, 12 neutrons), phosphorus-31 (15 protons, 16 neutrons) and gold-197 (79 protons, 118 neutrons).

Where elements of odd atomic number possess two naturally-occurring isotopes, in almost every case both have an even number of neutrons. Thus chlorine occurs as chlorine-35 and chlorine-37, with 17 protons and either 18 or 20 neutrons. Copper occurs as copper-63 and copper-65, with 29 protons and either 34 or 36 neutrons. Silver occurs as silver-117 and silver-119, with 47 protons and either 60 or 62 neutrons.

Elements of even atomic number, with three or more naturally-occurring isotopes, usually have a larger number with even numbers of neutrons than of odd (the latter being "even-odd nuclei"). As an example, xenon possesses nine naturally-occurring isotopes, of which seven are "even-even" (xenon-124, 126, 128, 130, 132, 134, and 136). The number of protons in each is 54, while the number of neutrons is 70, 72, 74, 76, 78, 80, and 82 respectively. Only two "even-odd" naturally-occurring xenon isotopes exist. These are xenon-129 and xenon-131, with neutron numbers of 75 and 77.

With one exception no element possesses more than two "even-odd" isotopes. The exception is tin, which contains three of them, tin-115, 117 and 119. Here the number of protons is 50, and the number of neutrons is 65, 67, and 69. (However, tin possesses seven "even-even" isotopes.)

The rarest of all nuclei are the "odd-odd nuclei" which contain odd numbers of both protons and neutrons. Only nine of these are naturally-occurring; of these nine, five are slightly radioactive and only the four simplest are completely stable.

The four stable "odd-odds" are hydrogen-2 (one proton, one neutron); lithium-6 (three protons, three neutrons); boron-10 (five protons, five neutrons); and nitrogen-14 (seven protons, seven neutrons). Of these, three are rare within their own element. Hydrogen-2 makes up only 1 out of 7000 hydrogen atoms; lithium-6 only 2 out of 27 lithium atoms; and boron-10 only 1 out of 5 boron atoms.

Nitrogen-14 is the surprising member of the group. It makes up 996 out of every 1000 nitrogen atoms, far outweighing the only other stable nitrogen isotope, nitrogen-15, an "odd-even" made up of seven protons and eight neutrons.

The alpha particle, made up of a pair of protons and a pair of neutrons, is particularly stable. When radioactive atoms eliminate nucleons, they never do so in units of less than an alpha particle.

The alpha particle is so stable that a nucleus made up of a pair of them (four protons plus four neutrons) is extremely unstable, almost as though the alpha particles are far too self-contained to have any capacity whatever to join together. Such a nucleus would be that of beryllium-8, which has a half-life of something like 3×10^{-16} seconds.

On the other hand, carbon-12, oxygen-16, neon-20, magnesium-24, silicon-28, sulfur-32, and calcium-40, which may be looked upon, after a fashion, as being made up of the union of 3, 4, 5, 6, 7, 8, and 10 alpha particles, respectively, are all particularly stable.

Some of the phenomena of natural radioactivity may be interpreted in the light of what has already been said. Atoms such as those of uranium-238 or thorium-232, in order to achieve stability, must reduce the number of protons in the nucleus to not more than 83.

To do so, alpha particles are ejected, but this eliminates neutrons as well as protons. When an equal number of protons and neutrons are eliminated, where neutrons are already present in excess, the n/p ratio rises. Thus the n/p ratio in uranium-238 (92 protons, 146 neutrons) is 1.59. If a uranium-238 managed to eject five alpha particles, it would lose ten protons and bring its atomic number down to 82 (that of lead) for possible stability. However, it would also have lost ten neutrons for a total decline in mass number of 20, and it would be lead-218. There the n/p ratio (82 protons, 136 neutrons) would be 1.66. So high an n/p ratio is completely incompatible with stability, and indeed lead-218 has never been detected. The most massive known lead isotope is lead-214, with a half-life of less than half an hour.

As the atomic weight is decreased, the n/p ratio must decrease also if stability is to be achieved. To do this, a neutron is converted to a proton and a beta particle is emitted. By a combination of alpha and beta emission, uranium-238 eventually becomes lead-206, with a loss of 10 protons and 22 neutrons and a decline in the n/p ratio from 1.59 to 1.51.

The regularities in proton-neutron combinations show clearly that stable nuclei are not built up in random fashion, but according to some orderly system. It has seemed to some physicists that just as orderliness was introduced into the chemical aspects of the elements by means of a periodic table eventually found to be based on electron shells (see Chapter 5), so order could be brought to the nuclear properties by means of a system of nuclear shells.

Such a system was advanced in 1948 by the Polish-American physicist Maria Goeppert-Mayer (1906–   ). She pointed out that isotopes containing certain numbers of protons or neutrons were particularly common or particularly stable. These numbers are called *shell numbers*, or, more dramatically, *magic numbers*, and they are 2, 8, 20, 50, 82, and 126.

Thus, helium-4 is made up of two protons and two neutrons; oxygen-16 of eight protons and eight neutrons and calcium-40 of 20 protons and 20 neutrons, and all three are particularly stable isotopes. Again, the element with the greatest number of stable isotopes is tin, whose nuclei contain 50 protons. There are also six naturally-occurring isotopes containing 50 neutrons (among which is rubidium-87, which is very slightly radioactive). There are seven stable isotopes containing 82 neutrons and four (those of lead) containing 82 protons.

Nor is it numbers of isotopes alone that count. Other nuclear properties seem to reach significant maxima or minima at the magic numbers. Thus isotopes containing a magic number of either protons or neutrons seem to have lower cross sections with respect to neutron absorption than do other isotopes of similar complexity.

Mrs. Goeppert-Mayer has attempted to account for these magic numbers by assuming the protons and neutrons to be arranged within the nucleus in *nucleon shells*, which are filled according to an arrangement of nuclear quantum numbers. The magic numbers are those at which key shells are completely filled (analogously to the situation of the inert gases in connection with electron shells).

This "nuclear periodic table" has had some triumphs. It has been used to predict which nuclides could exist in excited states for a significant length of time, forming nuclear isomers (see page 128). However, this model is still a subject of considerable controversy.

Packing Fractions

The stability of a particular nuclide rests not only on n/p ratios, but, more fundamentally, on the energy relationship of a particular nuclide with other nuclides of equal nucleon counts.

To see how this can be, let's begin by considering that though the mass number of an isotope is usually given as a whole number, it is not quite a whole number. We speak of oxygen-18, potassium-41 and uranium-235, assuming the mass numbers to be 18, 41, and 235 respectively.

Aston's mass spectrograph (see page 139) made it possible, however, to measure the mass of individual isotopes with great precision. We now know that on the carbon-12 scale the actual mass of the oxygen-18 nucleus is 17.99916, while that of potassium-41 is 40.96184 and that of uranium-235 is 235.0439.

This may seem strange in view of the belief that the nucleus is made up of protons and neutrons only, with each one of those particles possessing unit mass. But do protons and neutrons indeed have a mass of precisely 1? They do not. On the carbon-12 scale, the mass of the proton is 1.007825 and that of the neutron is 1.00865.

But this raises another question. The carbon-12 nucleus is made up of six protons and six neutrons. The 12 nucleons, considered one at a time, have a total mass of 12.098940, yet the mass of these same nucleons combined into a carbon-12 nucleus is 12.00000. There is a *mass defect* of 0.098940; what has happened to it?

In the light of Einstein's equation showing the equivalence of mass and energy (see page II–111), it seems clear that the extra mass has been turned into energy.

Six protons and six neutrons in combining to form a carbon-12 nucleus lose a little less than 1 percent of their total mass and liberate that as energy. If it were desired to break up the carbon-12 nucleus into individual nucleons again, that quantity of energy must be supplied it again. It is the difficulty of collecting and delivering this energy that keeps carbon-12 stable. The energy holding the nucleons within a nucleus together is much greater than the energy holding the atoms of a molecule, or the molecules of a solid mass together. It is also greater than the energy holding electrons within the atom. For that reason, procedures which suffice to melt a solid or decompose a compound, or even ionize an atom, fail utterly in any effort to break up an atomic nucleus.

Yet if nuclei cannot be broken up into individual nucleons

without prohibitive expenditure of energy, less drastic changes are possible; some of these less drastic changes even take place spontaneously.

To begin with, the more energy per particle given off in forming the nucleus by "packing together" individual nucleons, the more stable that nucleus tends to be (all other things being equal) A way of measuring this energy of nuclear formation is to subtract the mass number (A) from the actual mass of the isotope (A_m). This difference, the mass defect, can be divided by the actual mass to give the fractional mass defect. To remove the decimal, the result is customarily multiplied by 10,000 to yield what Aston called the *packing fraction*. If we let P_f represent the packing fraction, then we can say:

$$P_f = \frac{10,000 \, (A_m - A)}{A_m} \qquad \text{(Equation 11–1)}$$

The lower the packing fraction, the greater the loss of mass in forming the nucleus, and the greater the tendency toward stability.

Among the elements, the packing fraction is highest for hydrogen. The actual mass of the hydrogen-1 nucleus (the bare proton) is 1.007825. If this is substituted for A_m in Equation 11–1, and 1 is substituted for A, then the packing fraction turns out to be 78.25. This is not surprising since a single proton isn't

Aston's packing fraction curve

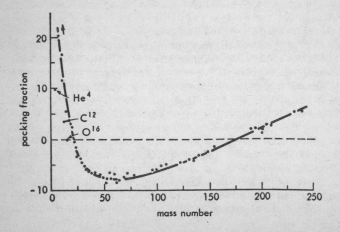

"packed together" at all in forming a nucleus. The packing fraction of a single neutron is even higher, for it is 86.7.

On the other hand, lithium-7, with A_m equal to 7.01601 and A equal to 7, has a packing fraction of 22.9, while carbon-13 with A_m equal to 13.00335 and A equal to 13, has a packing fraction of 2.4.

In general, starting at hydrogen-1, the packing fraction decreases for quite a while; this indicates that the nuclei of proper n/p ratio tend to grow more stable as they grow more complicated. To put it another way, if two very simple nuclei are combined to form a more complicated one, energy is released.

By the time nitrogen-15 is reached, the packing fraction is just about 0, but for still more complex nuclei the packing fraction falls into negative numbers. (This is a consequence of the fact we have chosen to let carbon-12 equal exactly 12. Had we established mass numbers on the basis of iron-56 equal to exactly 56, there would be no negative packing fraction numbers.)

For potassium-41, for instance, the value of A_m is 40.96184 and that of A is 41, so that the packing fraction is -9.3. A minimum is finally reached at iron-56, which has a packing fraction of -11.63. Thereafter, the packing fraction increases again, so that for tin-120, for instance, it is -8.1, and for iridium-191, -2.0. At the extreme end of the list of tables the packing fractions are positive again. For uranium-238, it is $+2.1$.

This means that the middle-sized atoms such as iron and nickel are the most stable of all. Not only is energy released if very simple atoms are built up to more complicated ones, but also if very complicated elements are broken down to less complicated ones.

All this is reflected in the general composition of the universe. On the whole, estimates of the distribution of elements in the universe, based on astronomic data, indicate that the more complicated an element, the rarer it is. Some 90 percent of the atoms in the universe are hydrogen (the simplest element) and another 9 percent are helium (the next simplest). Nevertheless, because of the particular stability of iron, it is to be expected that iron ought to be more common than other elements of similar complexity. This is indeed true, and our own planet, which is not massive enough to have retained the very simplest atoms, is about 35 percent iron in mass.

Packing fractions are particularly low for carbon-12 and oxygen-16 (which can be looked upon as being made up of alpha particles) and are especially low for helium-4 itself (which *is* the alpha particle). Thus the packing fraction of lithium-6 is 25.2 and that of hydrogen-2 is 70. Since helium-4 is midway between these in mass, one might suppose that its packing fraction would also fall roughly midway between. It is, however, only 6.5, much lower than either. It is not surprising then that helium, carbon and oxygen are among the more common atoms in the universe.

Whether a particular nuclide is stable or not, depends not only on its own packing fraction, but on the packing fraction of nuclides of equal nucleon number. For instance, sodium-24 (11 protons, 13 neutrons) might be suspected of stability if it could be considered by itself. However, magnesium-24 (12 protons, 12 neutrons) has a lower packing fraction. Therefore, if sodium-24 emits a beta particle and changes its nucleon arrangement from 11-13 to 12-12, it loses energy and gains stability. Beta emission is a cheap price to pay for this gain. Whereas it would take prohibitive energy to break up the sodium-24 nucleus altogether, only a slight amount of energy suffices to set off the change involved in a beta particle emission. Consequently, sodium-24 emits beta particles spontaneously and breaks down to magnesium-24, with a half-life of 15 hours.

Two neighboring isotopes of the same mass number cannot possibly both be stable. The one with the higher packing fraction spontaneously changes into the one with the lower. It's like rolling down an "energy hill" and the steeper the hill the shorter the half-life.

Two isotopes of the same mass number, but not neighbors, can both be stable. Thus, zinc-64 (30 protons, 34 neutrons) and nickel-64 (28 protons, 36 neutrons) are both stable, for they are separated by copper-64 (29 protons, 35 neutrons), which has a higher packing fraction than either. Both zinc-64 and nickel-64 may be visualized as occupying "energy hollows" with an "energy hump" (on which copper-64 is perched) lying between. Copper-64 is indeed unstable and can break down in either of two ways. It can emit a beta particle to become zinc-64, or it can emit the opposite of a beta particle (see page 223) to become nickel-64.

Sometimes the "energy hump" is very low, and the isotope existing there is nearly stable. This is the case of potassium-40 (19 protons, 21 neutrons), which lies between two stable isotopes, argon-40 (18 protons, 22 neutrons) and calcium-40 (20 protons, 20 neutrons). Potassium-40 is only slightly radioactive and it, too, breaks down in two fashions, one of which yields calcium-40 and the other argon-40.

Nuclear Energy

As soon as the existence of nuclear energy* was accepted, scientists began to speculate on the possibilities of putting it to use. Some isotopes already exist which, in effect, stand at the top of a very gentle slope and have been slowly rolling down it an atom at a time. These are the isotopes, uranium-238, uranium-235 and thorium-232, of course.

For instance, uranium-238 breaks down in a number of steps to form lead-206. In doing so, it gives off beta particles and gamma rays, the mass of which may be ignored, and also eight alpha particles, the mass of which may not be ignored. Including only the massive items, we can write: $U^{238} \longrightarrow Pb^{206} + 8He^4$.

The mass of the uranium-238 nucleus is 238.0506, that of the lead-206 nucleus is 205.9745, and that of the alpha particle is 4.00260. The total mass of the lead-206 nucleus plus eight alpha particles is 237.9953. This means that in the radioactive breakdown of uranium-238 to lead-206, each uranium-238 nucleus loses $238.0506 - 237.9953$, or 0.0553 atomic mass units.

This can be escalated to the gram level. If 238 grams of uranium break down completely to lead, then 55.3 milligrams of mass are converted to energy. This is a conversion of 0.225 milligrams of mass for each gram of uranium breaking down completely.

According to Einstein, $e = mc^2$, where e represents energy in ergs, m represents mass in grams, and c represents the velocity of light in centimeters per second. The velocity of light is 3×10^{10} centimeters per second, and squaring that gives us 9×10^{20}. If that is multiplied by 0.225 milligrams (or 2.25×10^{-4} grams) we find that a gram of uranium, breaking down completely, liberates 2.5×10^{17} ergs, or just under 5,000,000 kilocalories.

If one gram of gasoline is burned, some 12 kilocalories of energy are liberated. We see then that the energy delivered through

---

* Nuclear energy is commonly referred to as "atomic energy," and the phrase is even enshrined in such names as the "Atomic Energy Commission." This is certainly wrong, for the electrons are as much a. part of the atom as the nucleus is, and energy derived from chemical reactions, which involve electron transfers, have every right to be referred to as "atomic energy." Nevertheless, names such as "atomic energy" and equivalent misnomers such as "atomic bombs" and "atomic submarines" can never be wiped out for the more accurate "nuclear bombs" and "nuclear submarines." In this book, I shall use the adjective "nuclear" as a matter of principle and not because I expect it will change anything.

the radioactive breakdown of a gram of uranium is 420,000 times as great as that delivered by an equal mass of burning gasoline and is, in fact, equivalent to the explosion of about 5000 tons of TNT. This is a fair comparison of the relative intensities of nuclear energy and chemical energy.

But why, then, had man remained so unaware of this vast energy release by uranium? (He was always aware of the comparatively tiny energy release of, say, a candle flame.) The answer is clear. Uranium releases a great deal of energy in breaking down, but spreads that energy over a vast expanse of time. The gram of uranium delivers half its energy, or 2,500,000 kilocalories, over a period of 4,500,000,000 years. In one second it delivers far less energy than a candle flame does.

To be sure there are isotopes more intensely radioactive than uranium-238. Consider polonium-212, one of the daughter nuclides of thorium-232 and therefore always present in thorium ores. A gram of polonium-212 will decay to lead-208 by emitting one alpha particle. In doing so, it loses only 0.046 milligrams. This is only a sixth the mass lost by uranium-238 all-told, and so polonium-212 liberates only a sixth the energy. It liberates less than 1,000,000 kilocalories, or only about as much energy as would be turned loose by a mere 1000 tons of exploding TNT. However, polonium-212 has a half-life of less than a millionth of a second, and all that energy would be delivered in an instant. The explosion would be shattering. However, there is no way of accumulating a full gram of natural polonium-212. All earth's crust would have to be combed for it, and that still might prove insufficient.

After 1919, it became possible to deal with intensely radioactive nuclides without searching for them in nature; they could be synthesized. By bombardment with alpha particles or with artificially accelerated protons, stable nuclides could be knocked out of their "energy hollows," so to speak, and sent skittering up to some "energy hump" in the form of a radioisotope. The radioisotope would then, either quickly or slowly, slide back down into a hollow. Might not the energy released then be utilized?

Of course it might; and it is, every time a radioisotope is detected by a counter, and every time it is used as a source of bombarding particles. However, the amount of energy expended uselessly, in order to knock only an occasional atom out of its "energy hollow" is much greater than the amount that is gained when the resulting radioisotope rolls back.

How, then, arrange matters so as to turn a profit? For one thing, the return on the original outlay might be increased by making the exploding radioisotope itself do the work of forming more radioisotopes.

Thus, a carbon-12 nucleus, if struck by a neutron under the proper conditions, may absorb it and emit two neutrons. The reaction can be written:

$$_6C^{12} + _0n^1 \longrightarrow {}_6C^{11} + _0n^1 + _0n^1 \qquad \text{(Equation 11–2)}$$

or:

$$C^{12}(n,2n)C^{11}$$

Suppose then, a carbon-12 nucleus struck by a neutron gave up two neutrons, each of which struck a carbon-12 nucleus to produce a total of four neutrons, each of which . . . Nuclear reactions take place in millionths of seconds or less, so that if the number of breakdowns increases to 2, 4, 8, 16, 32 and so on in steps of millionths of seconds, the entire supply of carbon-12 would have undergone a nuclear reaction. Carbon-12, a very common substance, would deliver as much energy, as quickly, as polonium-212.

It is precisely this sort of thing that happens (on a chemical energy scale) when with a single match we burn down a forest. The match supplies the energy to set a leaf burning; the heat developed by the burning leaf ignites neighboring objects, and so on. Where the product of a reaction serves as the condition for continuing the reaction, chemists speak of a *chain reaction*. What the $(n,2n)$ process offered was the chance of a *nuclear chain reaction*.

However, the $(n,2n)$ process does not work. All the examples so far discovered require fast, energetic neutrons. A fast neutron sent into a carbon-12 target will liberate two neutrons, but slow ones. They are invariably much less energetic than the incident neutron; not energetic enough to initiate a new reaction of the sort that had given rise to them.

It is like trying to burn wet wood. You may set a tiny flame going, but it will not deliver enough heat to dry neighboring sections of wood so that they may burn—and the fire sputters out. Nor is this a bad thing. There are always stray neutrons present in the atmosphere and they are low energy. If they sufficed to start a nuclear chain reaction, then large parts of earth's crust might be subject to almost instantaneous nuclear explosions, and planets as we know them might not exist. The fact that the earth does exist may be evidence that no $(n,2n)$ process involving common atoms offers a practical nuclear chain reaction.

That was the situation, then, up to 1939. Although physicists knew that nuclear energy existed in tremendous quantities, there was no practical method of tapping it. There even seemed reason to believe that no practical method could conceivably exist. Rutherford, for example, was convinced (and said so) that the development of a practical source of large-scale nuclear energy was an idle dream. He died just a few years too soon to see his reasons refuted.

Nuclear Fission

The situation with respect to the utilization of nuclear energy changed radically in the late 1930's. Fermi had been bombarding uranium with thermal neutrons and had felt he had formed element 93. In a way he was right, but he had also induced other nuclear reactions that confused the results and left him puzzled.

Other physicists tackled the problem and were equally puzzled. Up to that point all nuclear reactions studied, whether natural or artificially produced, had involved the emission of small particles, no more massive than an alpha particle. Consequently, physicists tried to associate the various types of radioactivity in the bombarded uranium with atoms only slightly smaller than uranium.

The German physicist Otto Hahn (1879–   ) and his co-worker, the Austrian physicist Lise Meitner (1878–   ), found in 1938 that when barium compounds were added to the bombarded uranium, a certain type of radioactivity followed the barium through all the chemical manipulations to which it was subjected. Since barium is very like radium from the chemical standpoint (radium is just under barium in the periodic table), Hahn supposed he was dealing with a radium isotope.

However, nothing he could do would separate the barium carrier from the radium he supposed was accompanying it. Even manipulations that would ordinarily separate barium from radium failed. He found himself forced, little by little, to suppose that it was not a radium isotope he was dealing with, but a radioactive barium isotope.

Consider the consequences of this thought. The barium isotopes have an atomic number of 56, which is 32 less than that of uranium. To form a barium isotope, a uranium atom would have to unleash a flood of eight alpha particles, and no such flood of alpha particles was detected in neutron-bombarded uranium. It began to seem necessary to suppose that the uranium nucleus, upon absorbing a neutron, might simply break in half (more or less). This process came to be called *uranium fission* or, more generally, *nuclear fission,* since isotopes other than those of uranium were eventually found to be subject to it.

Such nuclear fission makes sense in that it involves a slide down an "energy hill" even more extensive than that brought about in ordinary radioactive transformations. Where under ordinary conditions uranium is converted to lead, with a lower packing fraction, in the case of fission, uranium is converted to such atoms as barium and krypton, which have still lower packing fractions.

Thus, while a gram of uranium, converted to lead by the ordinary radioactive process, will lose about 1/4 milligram, that same gram of uranium undergoing fission will lose just about 1 milligram in mass. In other words, uranium fission will yield some four times as much energy, gram for gram, as ordinary radioactivity will.

Uranium fission seems to fit in well with a theoretical model of nuclear structure advanced by Bohr. In this model, the nucleus is viewed as analogous to a drop of liquid (it is referred to, in fact, as the *liquid-drop model*). Instead of considering the nucleons as occupying different shells and behaving with relative independence, as in the shell model, the nucleons are considered as jostling each other randomly like molecules in a drop of liquid.

A neutron entering such a nucleus has its energy absorbed and distributed among all the nucleons very quickly, so that no one nucleon retains enough energy to eject itself from the nucleus. The surplus energy could be gotten rid of as a gamma ray, but there is also a distinct possibility that the entire nucleus will be set to oscillating as a liquid drop might under similar conditions. There is then a further possibility that before the energy could be eliminated as a gamma ray, the nucleus might oscillate strongly enough to break in two.

When a uranium nucleus breaks in two in this fashion, it does not always divide in exactly the same way. The packing fraction among nuclei of moderate size does not vary a great deal, and the nucleus may well break at one point in one case and at a slightly different point in another. For this reason, a great variety of radioisotopes are produced, depending on just how the division takes place. They are lumped together as *fission products.* The probabilities are highest that the division will be slightly unequal. with a more massive half in the mass number region of from 135 to 145 and a less massive half in the region of from 90 to 100.

It was among these fission products that isotopes of element number 61 were first isolated in 1948. The new element was named "promethium" because it was snatched out of the nuclear furnace in the same way that fire was supposed to have been snatched from the sun by the Greek demigod Prometheus.

In producing relatively small fission products, the uranium atom is brought to a portion of the list of elements where the n/p ratio is smaller. Fewer neutrons are needed in the nuclei of the fission products than in the original uranium nucleus, and these superfluous neutrons are liberated. In consequence, each uranium atom undergoing fission liberates two or three neutrons.

One might wonder why, if uranium fission liberates more energy than ordinary uranium breakdown does, uranium does not spontaneously undergo fission rather than ordinary breakdown. Apparently, before any change can take place, the nucleus must absorb a small quantity of energy that will carry it over an "energy hump" before it can start sliding down the "energy hill." We have here a sort of "nuclear ignition" that is analogous to the heat of friction that starts a match burning.

The higher the energy hump, the less likely an individual nucleus is to gain sufficient energy to pass over it in the ordinary course of events in which energy is constantly being randomly distributed and redistributed among subatomic particles. Therefore, the higher the energy hump, the fewer the nuclei that will undergo breakdown in any particular time interval, and the longer the half-life.

The energy hump is higher for uranium fission than for ordinary uranium breakdown, and it is therefore the latter which takes place and is detected, even though the former represents the greater overall stabilization.

Still, the energy hump for fission ought very occasionally to be overcome on a purely random basis (and not merely by the deliberate introduction of a neutron), and when that happens, the uranium nucleus ought to undergo fission without neutrons. Such *spontaneous fission* was discovered in 1940 by a pair of Russian physicists, G. N. Flerov and K. A. Petrjak.

Naturally, since fission has the higher energy hump, its half-life is longer. Whereas uranium-238 has an alpha emission half-life of some 4,500,000,000 years, it has a spontaneous fission half-life of some 1,000,000,000,000 years.

For the more massive transuranium isotopes, the spontaneous fission half-life decreases. For curium-242 it is a mere 7,200,000 years, and for californium-250 it is only 15,000 years.

12

Nuclear Reactors

Uranium-235

When Hahn first came to the conclusion that neutrons were initiating uranium fission, he hesitated to announce his finding since it seemed so "far out" a suggestion. At that time, however, Lise Meitner, his long time co-worker, being Jewish, had to flee Hitler's anti-Semitism and was in Stockholm.* The uncertainties of her own position made the risk of a "far out" scientific suggestion seem less dangerous, and she sent a letter to the scientific periodical *Nature*, discussing the possibility of uranium fission. The letter was dated January 16, 1939.

Niels Bohr learned of this by word of mouth and, on a visit to the United States to attend a conference of physicists, spread the news. The physicists scattered to see if they could confirm the suggestion. They promptly did so, and nuclear fission became the exciting discovery of the year.

To the Hungarian-American physicist Leo Szilard (1898–1964), what seemed most exciting (and unsettling) was the possibility of a nuclear chain reaction. He had been one of those who had considered the possibility before, and he had even

---

* Until 1938 she had been relatively safe, for she was an Austrian national, but in that year Hitler's Germany had forcibly annexed Austria.

patented a nuclear process that might possibly give rise to one (but didn't).

Uranium fission, however, offered a new approach. It was induced by slow thermal neutrons even more readily than by fast ones. The neutrons produced in the process of fission possessed ample energy for the induction of further nuclear fission. If anything, they had to be slowed down, which was easy.

World War II had started now, and Szilard, a refugee from Hitler's psychopathic tyranny, was fully aware of the terrible danger that faced the world if the Nazis tamed nuclear energy and put it to war use. Together with two other physicists of Hungarian birth, Eugene Paul Wigner (1902–   ) and Edward Teller (1908–   ), he set about interesting the American government in pursuing the project of developing methods for obtaining and controlling such a chain reaction.

They chose Albert Einstein as the only man with enough prestige to carry weight with non-scientists in such a matter. Overcoming Einstein's pacifistic scruples with difficulty, they persuaded the gentle physicist to write a letter on the subject to President Franklin D. Roosevelt. In 1941, Roosevelt was persuaded and he agreed to initiate a massive research program to develop a war weapon involving uranium fission. The final order was issued on December 6, the day before Pearl Harbor.

In order to establish a nuclear chain reaction, it is necessary to set up conditions radically different from those prevailing in the earth's crust. In the crust, although uranium is present and stray neutrons are to be found in the atmosphere, no chain reaction exists or, as far as we can tell, ever has existed.

The reason for this is that when an atom of uranium undergoes fission (either spontaneously or through absorption of a neutron), the neutrons liberated are absorbed by surrounding atoms. Most of these surrounding atoms are not uranium and are not themselves nudged into fission. The neutrons from fissioning uranium are thus absorbed and no neutrons are re-emitted, so that the potential chain reaction is effectively quenched. There is enough non-uranium material in even the richest natural concentration of uranium to quench any potential chain reaction at once.

What was necessary, then, if a nuclear chain reaction was to have any chance at all, was to make use of pure uranium, in the form of an oxide or even as the metal itself. In the metal, where almost all atoms would be uranium atoms, any neutron liberated by one uranium atom undergoing fission would stand an excellent chance of being absorbed by another uranium atom and therefore of bringing about another fission—the next link in the chain.

This, in itself, was a stiff requirement. In 1941, uranium had virtually no important uses, so that only small amounts of the metal were produced and those small amounts were not of high purity. Then, even as attempts to prepare large quantities of pure uranium began, an even more stringent qualification made itself evident.

Shortly after the idea of uranium fission had been accepted, Niels Bohr pointed out that on theoretical grounds uranium-235 was much more likely to undergo fission than uranium-238 was. Experiment soon showed Bohr to be right. This meant that ordinary uranium, even if highly purified, was still a poor material with which to set up a nuclear chain reaction, for 993 atoms out of every 1000 in such uranium would be uranium-238 which would absorb neutrons without undergoing fission—thus quenching the chain reaction.

To give a nuclear chain reaction a decent chance, uranium would have to be prepared in which uranium-235 was present in greater than usual amounts. Such a preparation would involve isotope separation, a difficult task—particularly, if it is to be carried through on a large scale.

Different isotopes of a given element have virtually identical chemical properties, and such differences as do exist depend on the fact that one isotope is more massive, and therefore reacts more sluggishly, than another. This difference is most marked in the case of hydrogen where hydrogen-2 is just twice as massive as hydrogen-1. This makes it possible for hydrogen-2 to be separated from hydrogen-1 with relative ease. The difference in mass between uranium-238 and uranium-235, however, is only 1.3 percent.

The best-established method for separating isotopes of small percentage difference in mass is by forcing a gas containing these isotopes as part of their molecules through some porous material (*diffusion*). The molecules must find their way through the pores, and those that contain less massive isotopes do so a bit more rapidly than do those which contain more massive ones.

The first samples of gas to emerge from the porous material are therefore "enriched" with a more than usual percentage of the light isotope, while the last samples to come through are "depleted" because they have smaller than usual percentage of the light isotope. The difference between the two fractions is very small, but the process may be repeated on each fraction. The smaller fractions can be recombined according to a fixed pattern and then separated again. Eventually, if this is continued long enough, the isotopes are nearly completely separated. The smaller the difference in mass between the isotopes, the greater the number of individual diffusions required.

Such a diffusion method requires a gas, of course, and neither uranium itself nor its most common compounds are gaseous. P. H. Abelson, however, suggested the use of uranium hexafluoride (UF_6) which, if not itself a gas at ordinary temperatures, is at least a volatile liquid with a boiling point at 56° C. It can therefore be maintained as a gas with little trouble.

The molecular weight of uranium hexafluoride containing uranium-238 is 352, while that of uranium hexafluoride containing uranium-235 is 349. The difference in molecular weight is only about 0.85 percent and diffusion had to be prolonged indeed to take advantage of so small a percentage mass difference. Giant installations (*diffusion cascades*) were set up at Oak Ridge, Tennessee, for this purpose in the early 1940's. In these, the UF_6 was put through numbers of porous barriers under conditions which automatically separated and recombined fractions in an appropriate manner. Eventually, enriched uranium hexafluoride was turned out at one end and depleted uranium hexafluoride at the other.

The "Atomic Pile"

Even as work on the purification of uranium and the separation of its isotopes proceeded, it was realized that a nuclear chain reaction could not, under even the best of conditions, be set up in a limited volume of uranium. Even uranium-235 atoms will not necessarily always absorb a neutron that comes blundering its way, toward the uranium atom. The neutron may merely bounce off, unabsorbed. It may do this over and over again, and it may be only the hundredth or the thousandth uranium-235 atom that will absorb it.

If in the process of bouncing from atom to atom the neutron manages to make its way out of the uranium and into the open air, it is lost. If enough neutrons do so, the nuclear chain reaction will be quenched. To prevent this, one must see to it that the chances of loss of neutrons to the surrounding environment, before absorption and consequent fission have a chance to take place, are minimized. The simplest way to do this is to increase the size of the uranium core in which fission is to take place. The larger its size the more bounces a neutron must undergo before reaching the edge of the core and the greater the chance of its absorption.

If the core is just large enough to lose so few neutrons that the nuclear chain reaction may just barely keep going, it is said to be at *critical size*. A smaller core, one of "subcritical size," cannot maintain a "self-sustaining nuclear reaction."

The critical size is not an absolute. It depends on the nature of the core, on its shape, and so on. A core of enriched uranium naturally has a smaller critical size than one of ordinary uranium, since the greater the concentration of uranium-235, the fewer the bounces required before absorption and the smaller the chance (at any particular core size) of escape into the air.

Then, too, the critical size can be reduced if slow neutrons, rather than fast neutrons, are used, since uranium-235 has a greater cross section for slow neutrons and fewer bounces will be required in their case. To slow the neutrons, a moderator (see page 174) is required, and very pure graphite serves the purpose well. Such a moderator could also serve as a neutron reflector. If the moderator is built about the uranium core, neutrons emerging from the core and striking the graphite will bounce here and there without being absorbed, and a number will bounce back into the uranium. In this way, the critical size is further reduced.

To control the nuclear chain reaction and prevent the uranium core from exploding, the reverse of a moderator is required. Instead of atoms that bounce the neutrons without absorbing them, as a moderator does, we need atoms that readily absorb the neutrons without either bouncing them or re-emitting them. Cadmium, some of the isotopes of which have a high cross section for neutrons, serves the purpose and "control rods" can be formed out of it.

Toward the end of 1942, the first attempt was made to set up a self-sustaining nuclear reaction. This took place under the guidance of Enrico Fermi (who had emigrated from Italy to the United States in 1938, but who was not yet an American citizen and was therefore technically an "enemy alien") under the stands of a football stadium at the University of Chicago.

At the time, some pure uranium was available in both metallic form and in the form of the oxide. It was not enriched and so the critical size was extraordinarily high. A very large "atomic pile" had to be built. (It was called a "pile" because it was, literally, a pile of bricks of uranium, uranium oxide, and graphite. In addition, "pile" was a neutral term that would not betray the actual nature of the structure if outsiders heard of it. After the war, "atomic pile" continued to be used for a short while and then gave way to a much more appropriate term, *nuclear reactor*.

When this first nuclear reactor was completed, it was 30 feet wide, 32 feet long and 21½ feet high. It weighed 1400 tons, of which 52 tons was uranium. The uranium, uranium oxide, and graphite were arranged in alternate layers with, here and there, holes into which long rods of cadmium could be fitted.

Suppose that in such a reaction a certain number of uranium atoms (n) undergo fission in a fixed unit of time, liberating x neutrons. Of these x neutrons, y do not find their mark but are absorbed by materials other than uranium, or are absorbed by

Oak Ridge Nuclear Reactor

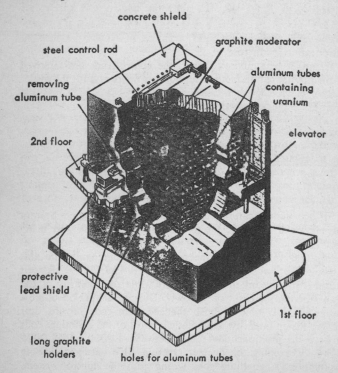

concrete shield

steel control rod

graphite moderator

removing aluminum tube

aluminum tubes containing uranium

2nd floor

elevator

protective lead shield

1st floor

long graphite holders

holes for aluminum tubes

uranium atoms which nevertheless do not undergo fission, or escape out of the reactor altogether. This means that *x-y* neutrons actually strike a uranium-235 atom and bring about fission. The ratio $(x-y)/n$ is the *multiplication factor*.

If the multiplication factor is less than 1, then at each succeeding link in the chain reaction, fewer atoms undergo fission and fewer neutrons are produced. The nuclear chain reaction is quickly quenched.

If the multiplication factor is greater than 1, then at each link in the chain a larger number of uranium atoms undergo fission and a greater number of neutrons are produced. In a fraction of a second, the intensity of the chain reaction escalates itself into a fearsome explosion.

In the reactor, as constructed at the University of Chicago, the multiplication factor was distinctly less than 1 with the cadmium control rods pushed all the way in. As the rods were slowly pulled out, less and less cadmium remained within the reactor to absorb neutrons, and more and more neutrons were consequently available to spur uranium atoms to fission. The multiplication factor rose.

It might be supposed that as the control rods are removed and the multiplication factor rises nothing happens until the multiplication factor edges the tiniest trifle over 1—at which time the entire pile explodes carrying part of the city of Chicago with it.

Fortunately, this need not happen. Almost all (but not quite all) the neutrons produced in the course of a nuclear chain reaction are produced virtually instantaneously as a uranium atom undergoes fission. These are *prompt neutrons*. About 0.75 percent of the neutrons, however, are produced by fission products and are emitted over a period of several minutes. These are *delayed neutrons*.

If the multiplication factor is above 1.0075, then prompt neutrons alone are sufficient to escalate the reaction and bring about an explosion at once. If the multiplication factor is between 1.0000 and 1.0075, the prompt neutrons cannot do this of themselves but must have the cooperation of the delayed neutrons. This means that for a short while the intensity of the fission reaction increases only slowly. During this period of slow increase there is time to push the cadmium rods inward thus reducing fission intensity. Automatic control of the cadmium rods can keep the multiplication factor between 1.0000 and 1.0075 indefinitely, keeping the nuclear reaction alive but not allowing an explosion. If anything goes wrong with the control system, matters are so arranged that the cadmium rods fall inward of themselves, quenching the reaction. This is a "fail-safe" situation, and a quarter-century of experience shows that nuclear reactors are quite safe when properly designed.

On December 2, 1942, at 3:45 P.M., the cadmium rods in Fermi's "atomic pile" were pulled out just enough to produce a self-sustaining reaction. That day and minute are taken to mark the beginning of the "atomic age." (Had the control rods been pulled out all the way, the multiplication factor would have been 1.0006—safe enough.)

News of this success was announced to Washington by the cautious telegram reading: "The Italian navigator has entered the new world." There came a questioning wire in return: "How were the natives?" The answer was sent off at once: "Very friendly."

The "Atomic Age"

Nuclear reactors have multiplied in number and in efficiency since Fermi's first "pile." Many nations now possess them, and they are used for a variety of purposes.

Neutrons are produced by the uranium atoms undergoing fission in unprecedented amount. They can be used to bombard a variety of targets and to produce radioisotopes in quantities that would be impossible under any other conditions. It is only since World War II, therefore, that radioisotopes have been available in relatively large quantity and at relatively low prices. Consequently, techniques involving such isotopes in biochemical research, in medicine and in industry have multiplied and flourished in the last few decades.

The nuclear reactor can also be used to produce power. The heat produced by the reactor can heat some high-boiling fluid passing through it (liquid sodium, for instance). This in turn can be used to boil water and form steam that will turn a turbine and produce electricity.

In 1954, the first nuclear submarine, the U.S.S. *Nautilus*, was launched by the United States. Its power was obtained entirely from a nuclear reactor, and it was not constrained to rise to the surface at short intervals to recharge batteries. Since it could remain underwater for extended periods of time, it was that much safer from enemy detection and attack.

The first American atomic surface ship was the N.S. *Savannah*, launched in 1959. Its nuclear reactors make use of enriched uranium dioxide as the fuel, and its 21 control rods contain neutron-absorbing boron.

In the mid-1950's, nuclear power stations were designed for the production of electricity for civilian use. The Soviet Union built a small station of this sort in 1954. It had a capacity of 50,000 kilowatts. The British built one of 92,000 kilowatt capacity which they called Calder Hall. The first American nuclear reactor for civilian purposes began operations at Shippingport, Pennsylvania, in 1958.

The greatest problem presented by such power stations (aside from expense, which may be expected to decrease as techniques grow more sophisticated) is the fact that the products of uranium fission are themselves radioactive.

What's more, as these fission products accumulate in the uranium core, they begin to interfere with operations. Some of them are relatively efficient absorbers of neutrons, so that they act to quench the nuclear chain reaction. Every two or three years, therefore, a nuclear reactor must be shut down (even though its fuel is very far from exhausted) and the fission products separated from the core.

The half-life of some of the fission products is 20 years or more, so it may be over a century before a batch of them can be considered no longer dangerously radioactive. For this reason, they must be disposed of with great care. Concentrated solutions can be encased in concrete and sealed in steel tanks, and then buried underground. Methods are also being investigated for fusing such fission products with silicates to form "glasses." This would be completely leak-proof and therefore safer to store.

The fission products themselves still contain energy, and some of them can be used in lightweight *nuclear batteries*. Such batteries are popularly termed SNAP ("Systems for Nuclear Auxiliary Power.") In such batteries, the heat given off by the radioactive breakdown of an isotope is used to raise the temperature of one end of a thermocouple (see page II–180) and produce electricity.

The first SNAP was constructed in 1956, and since then over a dozen varieties have been built. Some have been put to use in powering man-made satellites over long periods. SNAP batteries can be as light as four pounds, can deliver as much as 60 watts, and can have a lifetime of up to ten years.

Not just any radioisotope will do for nuclear batteries. It must have an appropriate half-life so that it will deliver heat neither too rapidly nor too slowly; it must be free of dangerous gamma ray emission, and be relatively cheap to prepare. Only a few radio-isotopes meet all necessary qualifications. The one most frequently used is the fission product strontium-90, which in another connection (see page 212) is one of the great new dangers to humanity.

The vision of a world in which uranium fission ekes out the energy supplies stored in coal and oil is dimmed somewhat by the fact that the prime nuclear fuel, uranium-235, is not extremely common. Uranium itself is not one of the rarest elements, but it is widely scattered throughout the earth's crust and concentrated pockets are rare. In addition, uranium-235 makes up only a small percentage of the metal.

Fortunately, uranium-235 is not the only isotope that can be stimulated into fission by neutron bombardment. Another isotope of this sort is plutonium-239. This does not exist in significant quantities in nature, but it can be formed by the neutron bombardment of uranium-238. This forms neptunium-239 first, then plutonium-239.

Once formed, plutonium-239 is easily handled, for it has a half-life of over 24,000 years. Therefore, in human terms its existence is just about permanent. Furthermore, it is not an isotope of uranium but a distinct element, so separating it from uranium is not nearly as difficult a problem as was isolating uranium-235.

During World War II, plutonium-239 was painstakingly gathered together so that its ability to undergo fission might be studied. A self-sustaining nuclear reaction can be maintained in plutonium-239 even under the impact of fast neutrons. Plutonium reactors (*fast reactors*) require no moderators and are therefore more compact than ordinary reactors.

Plutonium-239 can be produced as a by-product of power obtained from uranium-235. The neutrons emerging from a uranium-235 core can be used to bombard a shell of ordinary uranium surrounding the core. Quantities of uranium-238 in the shell are converted to plutonium-239. In the end, the quantity of fissionable material produced in the shell may actually be greater than that consumed in the core. This is a *breeder reactor*.

Such a breeder reactor makes uranium-238 indirectly available as a nuclear fuel and increases the fissionable resources of mankind over a hundredfold.

Another fissionable material is uranium-233, an isotope first discovered by Seaborg and his group in 1942. It is a daughter isotope of the neptunium series and therefore does not occur in nature. However, it has a half-life of 162,000 years, so once it is formed it can be handled without trouble.

When thorium-232 is exposed to neutron bombardment, it becomes thorium-233, which emits a beta particle with a half-life of 22 minutes to become protactinium-233. The latter, in turn, emits a beta particle with a half-life of 27 days to become uranium-233. Thus, if a thorium shell surrounds a nuclear reactor, fissionable uranium-233 can be formed within it and easily separated from the thorium. In this way, the earth's supply of thorium is added to its nuclear fuel potential.

Despite this enumeration of the peaceful aspects of nuclear fission, it must be remembered that the research project set up in 1941 had as its first purpose the development of an explosive weapon. What was wanted was a core that would exceed the multiplication factor by as much as possible. For this purpose, the critical mass must be as small as possible, since such a bomb ought to be transportable. Hence, pure uranium-235 or plutonium-239 ought to be used.

Such a bomb can safely be transported in two halves, since each portion would then be of subcritical size. At the crucial point, one half can be slammed against the other by means of an explosive. The stray neutrons in the air will suffice to build up an immediate nuclear explosion.

By 1945, uranium isotopes and plutonium had been prepared in sufficient quantity to construct three *fission bombs.** At 5:20 A.M. on July 16, 1945, at Alamogordo, New Mexico, one of them was exploded and was a horrifyingly complete success. The explosion had the force of 20,000 tons (20 kilotons) of TNT.

By that time, World War II was over in Europe, but not the war with Japan. It was decided to use the two remaining nuclear bombs against Japan. On August 6, 1945, one of them was exploded over the town of Hiroshima, and on August 8, the other was exploded over Nagasaki. Japan surrendered and World War II was over.

Nuclear Fusion

Even under the best of circumstances, energy drawn from fissionable fuel has its disadvantages. Taken together, uranium and thorium make up only about 1.2 parts per hundred thousand of the earth's crust. This represents, to be sure, perhaps ten times

* These are popularly termed "atomic bombs" or "A-bombs."

as much potential energy as can be obtained from the earth's total supply of coal, oil, and gas, but only a small part of earth's fissionable fuel supply can be extracted from the crust with reasonable ease. Then, too, even if all could be used, what would we do with the mounting accumulations of fission products—products impossible to keep and dangerous to dispose of?

A bright alternative offers itself at the other end of the packing fraction curve. Energy can be obtained not only by breaking down massive atoms into less massive ones, but by building up simple atoms into less simple ones. The latter situation is termed *nuclear fusion*.

The most obvious case is that in which hydrogen, the simplest atom, is fused to helium, the next simplest. Suppose, for instance, that we consider the following reaction:

$$_1H^2 + _1H^2 \longrightarrow _2He^4$$

The mass number of hydrogen-2 is 2.01410 and that of two of such nuclei is 4.02820. The mass number of helium-4 (which has an unusually low packing fraction) is 4.00280. The loss of mass is 0.0254 out of a total of 4.0282. The percentage loss of mass is 0.63, whereas in uranium fission it is only 0.056. In other words, on a weight for weight basis, over ten times as much energy is available in nuclear fusion as in nuclear fission.

Nuclear fusion brought itself to human attention in the sky first. In the mid-nineteenth century, when the law of conservation of energy was clearly established, physicists began to question the origin of the vast energies of the sun. The German physicist Hermann Ludwig Ferdinand von Helmholtz (1821–1894) had seen in the force of gravitation the only possible source of the sun's energy and had suggested that a slow contraction powered the solar radiation.

Unfortunately, with gravitation as the source of radiant energy, it seemed as though the earth could not have lasted more than a hundred million years or so. Prior to that, the sun would have had to be large enough to more than fill the earth's orbit; it would have had to be that large if enough contraction were to have taken place to support its radiation for a hundred million years.

Once radioactivity was discovered, however, it was possible to take a new look at the problem. The atomic nucleus was an energy source unknown to Helmholtz and to the men of his generation. It came to seem more and more reasonable to suppose that the sun's radiation was supported by nuclear reactions.

The nature of such a reaction, however, remained a puzzle for some decades. The earliest nuclear reactions known, those of uranium and thorium breakdown (or, for that matter, the later-discovered uranium fission) could not be useful in the sun, since there wasn't enough uranium, or massive atoms generally, in the solar sphere to supply the necessary energy under any circumstances.

Indeed, by atom count the sun was something like 85 percent hydrogen and 10 percent helium. It seemed quite likely, therefore, that if nuclear reactions powered the sun, they would have to be reactions involving hydrogen.

However, hydrogen does not, under conditions on earth, spontaneously participate in nuclear reactions. Conditions on the sun differ most dramatically with respect to temperature (the sun's surface is known to be at a temperature of 6000° C), but it was not at all certain that this difference was significant.

Early experiments with uranium and with other naturally radioactive elements made it appear that the radioactive process, unlike ordinary chemical reactions, was not affected by heat. The half-life of radium was not decreased by extreme cold or increased by extreme heat. Nor could two atoms which, at ordinary temperatures, did not engage in nuclear reactions, be made to do so by extreme heat.

Of course this depends on what is meant by "extreme heat." The temperatures available in the laboratories of the early twentieth century were insufficient to smash atoms together with such force as to break through the electronic "bumpers" and force nucleus against nucleus. Even the temperature of the solar surface was quite insufficient for the purpose.

However, the English astronomer Arthur Stanley Eddington (1882–1944) produced a convincing line of argument to show that if the sun were gaseous throughout, then it could be stable only if its interior temperature was extremely high—at millions of degrees.

At *such* temperature extremes, atomic nuclei could indeed be forced together, and nuclear reactions that would not take place at ordinary temperatures would become "spontaneous." A nuclear reaction that proceeds under the lash of such extreme heat intensities is termed a *thermonuclear reaction* ("thermo-" is from a Greek word for "heat"). Clearly, it must be thermonuclear reactions proceeding somewhere deep in the sun's interior that serve as the source of its radiant energy.

In 1938, the German-American physicist Hans Albrecht

Bethe (1906–   ) worked through the list of possible thermo-nuclear reactions involving the light elements, eliminating those that took place too quickly and would explode the sun, and those that took place too slowly and would let the sun's radiation die. The reaction upon which he finally settled began with the most likely candidate, the overwhelmingly present hydrogen.

He postulated that the hydrogen reacted with carbon to build up first nitrogen and then oxygen in a series of reactions. The oxygen atom broke up to helium and carbon. The carbon was thus ready to begin a new cycle and, since it was in the long run unchanged, behaved as a sort of "nuclear catalyst." The net effect of the series of reactions was to convert hydrogen-1 to helium-4. In later years, other sets of reactions involving more direct changes of hydrogen-1 to helium-4 were also proposed.*

The energy released by such nuclear fusion of hydrogen to helium (with or without a carbon catalyst) is quite sufficient to maintain the sun's radiation. The energy is obtained, of course, at the expense of the sun's mass. In order to keep its radiation going at the observed rate, the sun must lose 4,600,000 tons of mass every second. To do that, it must convert 650,000,000 tons of hydrogen-1 to helium-4 every second. However, the sun has so huge a supply of hydrogen-1 that although it has been radiating for five or six billion years there is enough left to power it for billions of years more.

With the development of the fission bomb, scientists had a method of achieving, even if only momentarily, temperatures high enough to bring about nuclear fusion here on earth. The dreadful power of such a weapon (a *fusion bomb*) was such that numerous scientists hesitated to proceed in that direction. Among those who hesitated was Oppenheimer, who in 1954 was to pay for this by suffering a form of political-scientific disgrace when his access to secret information was withdrawn. Prominent among those who condemned Oppenheimer and pushed for the development of the fusion bomb was Edward Teller, whose contribution to the problem was such that he later received the rather unenviable distinction of being called the "father of the hydrogen bomb."

In 1952, the first "thermonuclear device," or "hydrogen bomb," or "H-bomb" (for the fusion bomb is known by all these names) was exploded by the United States in the Marshall Islands. It was not long after that, that the Soviet Union developed its

---

* The details of these reactions—together with those reactions that involve helium fusion, and so on, during the later stages of a star's life-cycle—are more fittingly discussed in a book on astronomy.

own fusion bomb, and that, later still, Great Britain became the third thermonuclear power. (France and China, which have exploded fission bombs, have not yet developed fusion bombs.)

Where the first fission bomb had the explosive force of 20,000 tons of TNT, fusion bombs with an explosive force of 50,000,000 tons of TNT (50 megatons) and beyond have been exploded.

Radiation Sickness

The fusion bomb escalated the danger to humanity (and to life on earth generally) by several notches. It was not merely that the explosion was far worse than that of the fission bomb, but rather that the long-lived effects of the products of a fusion bomb (even one exploded experimentally in peacetime) were insidious in the extreme. This is owing to the effect of high-energy radiation on living tissue.

Soon after the discovery of X rays, it was discovered that overexposure to such radiation gave rise to skin inflammations and burns that healed very slowly. The same proved to be true of the radiations from radioactive substances. Pierre Curie deliberately exposed himself to such radiations and reported the lingering symptoms that resulted.

The energy of X rays, gamma rays or speeding subatomic particles is sufficient, if absorbed by a molecule, to break chemical bonds with the production of high-energy molecular fragments (*free radicals*). These will, in turn, react with other compounds. A subatomic particle that is absorbed by an atom may alter its nature and, therefore, that of the molecule of which it is part. If the new atom is radioactive and emits a particle, the recoil will rupture the molecule even if it had survived intact till then.

Such chemical changes may well disrupt the intricately interrelated chemical machinery of a cell and upset those systems of reactions that control cellular cooperation. Changes may be induced, for instance, which will allow the unrestrained growth of certain cells at the expense of their neighbors and cancer will result. The skin, which bears the brunt of the onslaught of radiation, and those portions of the body such as the lymphoid tissue and the bone marrow, which produce blood cells, are particularly subject to this. (Even excessive exposure to ultra violet radiation increases the likelihood of the development of skin cancer.)

Leukemia, an unrestrained production of white blood cells (a condition which is slowly, but invariably, fatal) is one of the more likely results of excessive exposure to radiation. Both Marie

Curie and her daughter, Irène Joliot-Curie, died of leukemia, presumably as a result of longtime exposure to the radiation of radioactive substances.

Where radiation exposure is particularly great, enough destruction is wrought among the particularly sensitive tissues to break down cellular chemistry completely and bring about death in a period of weeks or months. Such *radiation sickness* was studied on a large scale for the first time among the survivors of the fission bombing of Hiroshima and Nagasaki.

Even worse than the death, fast or slow, of the individual is the long-term danger that continues over the generations. An altered molecule may not very seriously affect the individual within which it exists, for it will be present in only a few cells; however, it may be transmitted to a child born of that individual, and that child may have the altered molecule in every cell. The child will have undergone a *mutation*.

Mutations may also take place spontaneously, as the result of the natural radiation arising from radioactive substances in the soil, and natural radiation arising in outer space, as well as the result of random imperfections in the reproduction of key molecules. The rate of mutation will increase, however, as the general radiation from the environment rises because of nuclear bombs. Such mutations are generally for the worse and if produced at too great a rate will swamp the human species with a "mutation load" too great for safety.

Attempts have therefore been made to determine what amount of radiation can reasonably be borne by individuals (and by mankind in general) without making the danger acute.

A unit of radiation is the *roentgen*, abbreviated as *r*, named in honor of the discoverer of the X rays. This is defined as the quantity of X rays or gamma rays required to produce a number of ions equivalent to 1 electrostatic unit of charge (see page II–164) in a cubic centimeter of dry air at 0° C and 1 atmosphere pressure. (For this, a little over two billion ions of either sign must be formed.)

This unit applied originally to energetic electromagnetic radiation only. However, energetic particles produce the same sort of symptoms and effects that radiation does, and an effort was made to apply the unit to these particles. A *roentgen equivalent physical*, or *rep*, was spoken of as that quantity of radiation of particles which, on absorption by living tissue, produce the same effect as the absorption of 1 r of X rays or gamma rays.

The same effect is not always produced by a given quantity

of a given radiation on all living species. If one wishes to specify the effect on man, one speaks of a *roentgen equivalent man*, or *rem*, as that quantity of radiation of particles which, on absorption by the tissues of a living man, produces the same effect as the absorption of 1 r of X rays or gamma rays.

Massive particles are particularly dangerous to man. Thus, 1 r of X rays, gamma rays or beta particles can also be expressed as 1 rem. However 1 r of alpha particles must be expressed as 10 to 20 rem. In other words, the absorption of alpha particles is at least ten times as dangerous to man as the absorption of the same amount of ionizing potential in the form of beta particles.

The roentgen and the units derived from it are unsatisfactory in some respects because they measure ion-production, and the quantity of energy required to form ion-pairs in the case of some types of radiation can be a rather complicated quantity to determine. Therefore, the *rad* (short for "radiation") was introduced and has grown popular. This is a direct measure of energy. One rad is equivalent to the absorption of enough radiation in any form to liberate 100 ergs of energy per gram of absorbing material. Under most cases, the rad is just about equal to the roentgen.

Background radiation is the unavoidable radiation arising from outer space, from radioactive substances in the soil, and so on. It is estimated that the average human being receives 0.050 rem per year from radiation from outer space and another 0.050 rem per year from the natural radioactivity of the soil. In addition, there is 0.025 rem per year from the body's own radioactivity in the form of potassium-40 and carbon-14. The total background radiation is thus about 0.125 rem per year, and this must be consistent with life since we are all subjected to it and life generally has been subjected to it from the beginning. Indeed, in parts of the world in which radioactivity is higher than average and in which high altitudes and high latitudes combine to make radiation from space more intense, background radiation of as much as 12 rem per year have been reported.

Obviously, experimentation to see how high a level of background radiation can be tolerated is unthinkable, but experts in the field had estimated that general body exposure to 500 rem per year is tolerable. Those working with radioactive materials are guarded against absorbing more than some safe limit of radiation each week (it is supposed that absorption at a higher rate than 500 rem per year is tolerable if it is for short periods or is localized to parts of the body). For instance, badges may be worn containing strips of film behind various filters that will be penetrated only by the sort of energetic radiation being guarded against. The extent to which the film blackens will measure the extent of exposure.

Exposure to 100 r over a few days will kill most mammals, but it takes several million r to sterilize food completely by killing all the microorganisms. All the nuclear testing so far performed has not brought the radiation level anywhere near such lethal levels and indeed has not even contributed more than a comparatively small fraction of the background radiation that already exists.

However, every little bit hurts, and it was the general disapproval by public opinion that finally forced the thermonuclear powers to agree to a ban on such nuclear bomb testing as would increase the radiation level.

It was the fusion bomb that made the danger intense. The fission fragments produced by fission bombs are spread only locally and present only a limited danger (however horrible within that limit). The far greater force of the fusion bomb, however, lifts the fragments of its fission-trigger high into the stratosphere, where they may circulate for a period of years and then slowly settle over the world generally. It is the danger of this stratospheric *fallout* (a word coined in 1945, after the first nuclear explosions) that presents mankind with its greatest radiation hazard.

That the fallout danger is real was made plain at once. The first large fusion bomb, exploded in the Marshall Islands on March 1, 1954, contaminated 7000 square miles with radiation.

Among the more dangerous fission fragments are strontium-90 and cesium-137. Strontium-90 has a half-life of 28 years, so it remains dangerously radioactive for a century and more. Because of the chemical similarity of strontium to calcium, strontium-90 is concentrated in the calcium-rich milk of mammals feeding on strontium-90 contaminated vegetation. Children drinking such contaminated milk then concentrate the strontium-90 in their calcium-rich bones. The atomic turnover in the bones is relatively slow, and therefore the *biologic half-life* of strontium-90 is long. (That is, it takes a long time for the body to remove half of what it has absorbed, even after it is protected against further absorption.) In the bones, moreover, strontium-90 is in dangerously close contact with sensitive blood-cell forming tissues.

Cesium-137, with a half-life of 30 years, is another dangerous fragment. It remains in the soft tissues and has a shorter biologic half-life. However, it emits energetic gamma rays and, while in the body, can do significant damage.

Fusion Power

Naturally, it is not only for the sake of its destructive potentialities that fusion processes are of interest. If nuclear fusion could be made to proceed at a controlled pace, the energy requirements of mankind would be solved for the foreseeable future.

The advantage of fusion over fission involves first the matter of fuel. Where the fission fuels are comparatively rare metals, uranium and thorium, the fusion fuel is a much more common and readily available element, hydrogen. It would be most convenient if it were hydrogen-1 that were the specific isotope suitable for man-made fusion since that is the most common form of hydrogen. Unfortunately, the temperatures required for hydrogen-1 fusion, at a rate fast enough to be useful, are prohibitively high. Even at the temperatures of the solar interior, hydrogen-1 undergoes fusion slowly. It is only because of the vast quantity of hydrogen-1 available in the sun that the small percentage that does fuse is sufficient to keep the sun radiating as it does. (To be sure, if hydrogen-1 were more readily fusible than it is, the sun—and other stars—would explode.)

Hydrogen-2 (deuterium) can be made to undergo fusion at a lower temperature, and hydrogen-3 at a lower temperature still. However, hydrogen-3 is unstable and would be extremely difficult to collect in reasonable quantities. That leaves hydrogen-2 as the best possible fuel.

Two atoms of deuterium can fuse in one of two ways with equal probability:

$$H^2 + H^2 \longrightarrow He^3 + n^1$$

and:

$$H^2 + H^2 \longrightarrow H^3 + H^1$$

In the latter case, the H^3 formed reacts quickly with another H^2, thus:

$$H^3 + H^2 \longrightarrow He^4 + n^1$$

The overall reaction, then, would be:

$$5 H^2 \longrightarrow He^3 + He^4 + H^1 + 2 n^1$$

The energy produced from such a fusion of five deuterium atoms (let's call it a "deuterium quintet") is 24.8 Mev. Since 1 Mev is equivalent to 1.6×10^{-6} ergs, the deuterium quintet, on fusion, yields 4.0×10^{-5} ergs.

A gram-molecular weight of hydrogen-2 contains 6.023×10^{23} atoms. Since a gram-molecular weight of hydrogen-2 is two grams, one gram of hydrogen-2 contains 3.012×10^{23} atoms. Dividing this figure by five, we find that a gram of hydrogen-2 contains 6.023×10^{22} deuterium quintets. The total energy produced by the complete fusion of one gram of hydrogen-2 is therefore 2.4×10^{18} ergs. Since there are 4.186×10^{10} ergs to a kilocalorie, we can say that the complete fusion of one gram of hydrogen-2 produces 5.7×10^{7} kilocalories.

To be sure, only 1 out of every 7000 hydrogen atoms is hydrogen-2. Allowing for the fact that that one atom is twice as massive as the remaining 6999, we can say that one liter of water weighs 1000 grams, that 125 grams of it are hydrogen, and that of that hydrogen 43 milligrams are hydrogen-2. We can therefore say that the complete fusion of the hydrogen-2 contained in a liter of water will yield about 2.5×10^{6} kilocalories.

This means that by the fusion of the hydrogen-2 contained in a liter of ordinary water, we would obtain as much energy as we would get through the combustion of 300 liters of gasoline.

Considering the vastness of the earth's ocean (from all of which hydrogen-2 is easily obtainable) we can see that the earth's supply of hydrogen-2 is something like 50,000 cubic miles. The energy that could be derived from this vast volume of hydrogen-2 is equivalent to the burning of a quantity of gasoline some 450 times the volume of the entire earth.

Obviously, if fusion power could be safely and practically tapped, mankind would have at its disposal an energy supply that should last for many millions of years. And to top off that joyful prospect, the products of the fusion reaction are hydrogen-1, helium-3, and helium-4, all of which are stable and safe, plus some neutrons which could be easily absorbed.

There is one catch to this prospect of paradise. In order to ignite a hydrogen-2 fusion reaction, a temperature of the order of $100,000,000°$ C must be reached. This is far higher than the temperature of the solar interior, which is only $15,000,000°$ C, but then the sun has the advantage of keeping its hydrogen under enormous pressures, pressures unattainable on the earth.

Any gas at such a temperature on earth would, if left to itself, simply expand to an excessively thin vapor and cool almost instantaneously. That this does not happen to the sun is due to the sun's mass, which produces a gravitational field capable of holding gases together even at the temperature reached in the solar interior.

Such gravitational fields cannot be produced on earth, of course, and the hot gas must be kept in place some other way. Material confinement would seem to be out of the question, for a hot gas making contact with a cool container would cool off at once—or heat the container itself to a thin gas. A gas cannot be both hot enough for fusion and contained within a solid substance.

Fortunately, another method offers itself. As the temperature rises, all atoms are progressively stripped of their electrons and all that then exists are charged particles, negatively-charged electrons plus positively-charged nuclei. Substances made up of electrically charged atom-fragments, rather than intact atoms, are called *plasma*.

Investigators grew interested in *plasma physics* chiefly as a result of interest in controlled fusion, but, by hindsight, we now see that most of the universe is plasma. The stars are plasma, and here on earth phenomena such as ball lightning are isolated bits of plasma that have achieved temporary stability. Plasma even exists in man-made devices—for instance, within neon light tubes.

Plasma, consisting as it does of charged particles, can be confined by a nonmaterial container, a properly shaped magnetic field. The effort of physicists is now engaged in attempting to design magnetic fields that will keep plasma stably confined for periods long enough to initiate a fusion reaction—and to make the plasma hot enough for the fusion reaction to ignite. It is estimated that at the critical point, using gas, which at ordinary temperatures would be only 1/100 or less the density of the atmosphere, the pressures which would have to be withstood by the magnetic field at the point of fusion ignition would be something like 1500 pounds per square inch, or 100 atmospheres.

The requirements are stringent, and after a decade of research, success still lies frustratingly beyond the fingertips. Temperatures of about 20,000,000° C have been attained. Magnetic fields capable of containing the necessary pressures have been produced. Unfortunately, the combined temperature and pressure can be maintained only for millionths of a second, and it is estimated that at least a tenth of a second duration must be obtained in order for the first man-made controlled fusion reaction to be produced.

There is nothing (as far as we know) but time and effort standing in the way.

CHAPTER 13

Anti-Particles

Cosmic Rays

So far, we have populated our atomic world with nothing more than electrons, protons and neutrons, and yet have managed to explain a great deal. In the early 1930's, these subatomic particles were the only ones known and it was rather hoped they would suffice, for there would then be an agreeable simplicity to the universe. However, some theoreticians were pointing out the necessity for further types of subatomic particles, and the first discoveries of such particles arose out of the vast energies present in radiation bombarding earth from outer space. It is to this radiation that we will turn now.

As the twentieth century opened, physicists were on the watch for new forms of radiation. The coming of radio waves, X rays and the various radioactive radiations had sensitized them to such phenomena, so to speak.*

---

* The sensitization was too great in one case. In 1903, a reputable French physicist, Prosper Blondlot, reported the existence of a new type of radiation from metal under strain. He and others published many reports on this radiation, which Blondlot termed "N rays," the N standing for Nancy, the French city in which he held his university appointment. There seems no question but that Blondlot was utterly sincere. Nevertheless, the N rays were an illusion, his reports proved worthless, and his scientific career was blasted. The story is important if only to demonstrate that scientists are not infallible and that "scientific evidence" is not necessarily trustworthy.

Nevertheless, the most remarkable discovery of this sort arose out of the attempt to exclude radiation rather than to detect it. The gold-leaf electroscope, which was early used to detect penetrating radiation (see page 109), worked too well. A number of investigators, notably C. T. R. Wilson, of cloud chamber fame, had reported, by 1900, that the electroscope slowly lost its charge even when there were no known radioactive materials in the vicinity. Presumably, most reasoned, the earth's crust was permeated with small quantities of radioactive materials everywhere, so that stray radiation was always present.

Yet other investigators found that even when the electroscope was taken out over stretches of water remote from land, or better yet, when it was shielded by a metal opaque to known radiation and producing no perceptible radiation of its own, the loss of charge on the part of the electroscope was merely diminished. It did not disappear.

Finally, in 1911, the Austrian physicist Victor Franz Hess (1883–1964) took the crucial step of carrying an electroscope up in a balloon, in order that several miles of atmosphere might serve as the shield between the earth's slightly radioactive crust and the charged gold leaf. To his surprise, the rate of discharge of the electroscope did not cease; instead it increased sharply. Later balloon flights confirmed this, and Hess declared that the radiation, whatever it was, did not originate on earth at all, but in outer space.

Robert Millikan (who measured the charge on an electron), took a leading part in the early investigations of this new radiation, and suggested in 1925 that it be named *cosmic rays*, because this radiation seemed to originate in the cosmos generally.

Cosmic rays are more penetrating than either X rays or gamma rays, and Millikan maintained that they were a form of electromagnetic radiation even shorter in wavelength and higher in frequency than gamma rays. Nevertheless, as in the case of X rays and gamma rays, many physicists suspected that the radiation might be particulate in nature.

In this case, since the radiation came from outer space, a method of distinguishing between electromagnetic radiation and particles offered itself. If the cosmic rays were electromagnetic radiation, they would fall on all parts of earth's surface equally, assuming that they originated from all directions. They would not be affected by the earth's magnetic field.

If, on the other hand, they were charged particles, they would be deflected by earth's magnetic lines of force, those particles lower in energy being the more deflected. In this case, cosmic rays would be expected to concentrate toward earth's magnetic poles and to strike earth's surface with least frequency in the vicinity of its magnetic equator.

This *latitude effect* was searched for through the 1920's, particularly by the American physicist A. H. Compton (1892–1962). By the early 1930's, he was able to show that such a latitude effect did exist and that cosmic rays were particulate and not electromagnetic. One might therefore refer to *cosmic particles*.

The Italian physicist Bruno Rossi (1905–   ) pointed out, in 1930, that if the cosmic rays were particulate in nature, earth's magnetic field ought to deflect them eastward if the particles were positively charged, so that more of them would seem to be coming from the west than from the east. The reverse would be true if the particles were negatively charged.

To detect such an effect, it was insufficient merely to detect the arrival of a cosmic particle; one had to tell the direction from which it had arrived. To do this, use was made of a *coincidence counter*, which had first been devised by the German physicist Walther Bothe (1891–1957). This consisted of two or more G-M counters placed along a common axis. An energetic particle would pass through all of them, provided it came along that axis. The electric circuit was so arranged that only the discharge of all the counters at once (and an energetic particle passes through all the counters with so little a time interval between that it may be considered a simultaneous discharge) will register and be counted. The counters can be oriented in different directions to form a "cosmic-ray telescope."

By placing a cloud chamber among the counters, one can arrange the circuit so that the chamber is automatically expanded when the counters discharge. The ions linger a short interval and are caught by the droplets formed by the expanding cloud chamber. If a camera is also rigged to take photographs automatically as the chamber expands, the cosmic particle ends by taking its own picture.

Using coincidence counters, the American physicist Thomas Hope Johnson (1899–   ) was able to show, in 1935, that more cosmic particles approached from the west than from the east. Thus it was determined that cosmic particles were positively charged.

An understanding of the actual nature of the cosmic particles was hampered by the fact that many did not survive to reach earth's surface. Instead, they struck one or another of the atomic nuclei present in the atmosphere, inducing nuclear reactions and producing a highly energetic *secondary radiation*. Some of this secondary radiation consists of neutrons which can, in turn, react with nitrogen-14 to produce carbon-14 in an (n,p) reaction. Or it can knock a triton (H^3) out of a nitrogen-14 nucleus, producing carbon-12 in an (n,t) reaction. These tritons are the source of the small quantities of H^3 existing on earth.

Cosmic particles can produce other events that cannot easily be duplicated in the laboratory simply because even now we have no way of producing particles with the energy of the most penetrating of the particles from outer space. Where man-made accelerators can now produce particles with energies of 30 Bev or more, cosmic particles with energies in the billions of Bev have been recorded.

Such super-energetic particles possess these energies partly because they are massive and partly because their velocities are great—nearly as high as the ultimate velocity, that of light in a vacuum. When such extremely rapid particles burst through transparent matter (water, mica, glass), they are scarcely slowed. Light itself, however, is slowed down appreciably in these substances, in inverse proportion to the index of refraction (see page II–25). It may follow, then, that within some forms of matter a charged particle may travel considerably faster than light does in that form of matter (but never faster than light does in a vacuum).

Such a "faster-than-light" particle throws back light radiation in a sort of shock effect, analogous to the manner in which a faster-than-sound bullet throws back a cone of sound waves. This effect was first noticed by the Russian physicist Pavel Alekseyevich Cerenkov (1904–) in 1934, and it is therefore called *Cerenkov radiation*.

The wavelength of the Cerenkov radiation, its brightness, and the angle at which it is emitted can all be used to determine the mass, charge and velocity of the moving particle. Following a suggestion in the late 1940's by the American physicist Ivan Alexander Getting (1912–), *Cerenkov counters* were developed that react to the radiation and thus distinguish very energetic particles from among floods of ordinary ones, and serve also as source for much information about the former.

The late 1940's also saw the beginning of investigations of radiation by high-altitude balloons and by rockets. At elevated altitudes, the *primary radiation*—the original cosmic particles, and not those produced by collisions of those particles with nuclei—could be detected. It turned out that the large majority (roughly 80 percent) of the cosmic particles were very energetic protons and most of the remainder were alpha particles. About 2.5 percent of the particles were still heavier nuclei, ranging up to iron nuclei.

It very much seemed as though the cosmic particles were the basic material of the universe, stripped down to bare nuclei. The proportion of the elements represented was very much like that in typical stars such as our sun.

In fact, the sun is at least one source of cosmic particles. A large solar flare will give rise, shortly afterward, to a burst of cosmic particles falling upon the earth. Nevertheless, even though the sun is one source, it can't be the only one, or even a major one, as otherwise the direction from which cosmic particles arrive would vary markedly with the position of the sun in the sky—which it doesn't. Moreover cosmic particles from the sun are comparatively low in energy.

This raises the question: How do cosmic particles gain their tremendous energies? No nuclear reactions are known which would supply sufficient energy for the more energetic cosmic particles. Even the complete conversion of mass into energy would not turn the trick.

It seems necessary to suppose that the cosmic particles are, at the start, protons and other nuclei of high, but not unusually high, energy. They are then accelerated in some natural accelerator on a cosmic scale. The magnetic fields associated with the sun's spots might accelerate such particles to moderate energies. More energetic particles might have been produced by stars with more intense magnetic fields than our sun, or even by the magnetic field associated with the Galaxy as a whole.

The Galaxy, in this respect, might be looked upon as a gigantic cyclotron in which protons, and atomic nuclei generally, whirl about, gaining energy and moving in a constantly widening spiral. If they do not collide with some material body for a long enough time, they eventually gain enough energy to go shooting out of the Galaxy altogether.

The earth interrupts the flight of these particles (in all directions) at all stages of their energy-gaining lifetime. The most energetic particles may be those that went shooting out of some other galaxy at their energy peak. It may be that some galaxies, with unusually intense magnetic fields, may accelerate cosmic particles to far greater energies than our Galaxy does and may be important sources for these most energetic particles. Such "cosmic galaxies" have not yet been pinpointed.

The Positron

Now let us look at the list of particles known in the early 1930's when the nature of cosmic radiation was first being unraveled. There are the proton, neutron and electron, of course. In addition, there is a massless "particle," the photon, which is associated with electromagnetic radiation.

The photon makes it unnecessary to rely on the notion of action at a distance in connection with electromagnetic phenomena (see page II–139), and that can give rise to speculation concerning that other long-distance phenomenon, gravitation.

Some physicists suggest that gravitational effects also involve the emission and absorption of particles, and the name *graviton* has been given to these particles. Like the photons, these particles are visualized as massless objects that must therefore (like all massless particles) travel at the velocity of light.

Gravitation is an incredibly weak force, however. The electrostatic attraction between a proton and an electron, for instance, is about 10^{40} times as strong as the gravitational attraction between them. The graviton must therefore be correspondingly weaker than the average photon—so weak that it has never been detected and, as nearly as can be told now, is not likely to be detected in the foreseeable future. Nevertheless, to suppose its existence rounds out the picture of the universe and helps make it whole.

We can now list the five particles in Table XIII and include some of the properties determined for them. (Those for the graviton are predicted and not, of course, observed.)

In the 1950's, the custom arose of lumping the light particles

TABLE XIII—*Some Subatomic Particles*

Particle	Symbol	Mass (electron = 1)	Spin (photon = 1)	Electric Charge	Half-Life (seconds)
graviton	g	0	2	0	stable
photon	γ	0	1	0	stable
electron	e	1	½	− 1	stable
proton	p	1836	½	+ 1	stable
neutron	n	1839	½	0	1013

together as *leptons* (from a Greek word meaning "small") and the heavy particles as *baryons* (from a Greek word meaning "heavy"). Using this classification, the graviton, photon and electron are leptons, while the proton and neutron are baryons.

It would seem quite neat if these three leptons and two baryons represented all that existed in the universe—both matter and energy—and that out of them were built the hundred-odd atoms—out of which, in turn, all the manifestations of the universe from a star to a human brain were constructed.

The first indication that mankind was not to rest in this Eden of simplicity came even before the neutron was discovered. In 1930, the English physicist Paul Adrien Maurice Dirac (1902–), working out a theoretical treatment of the electron, showed that it ought to be able to exist in either of two different energy states. In one of those energy states, it was the ordinary electron; in the other, it carried a positive charge rather than a negative one.

For a while, however, this remained a theoretical suggestion only. In 1932, however, the American physicist Carl David Anderson (1905–   ) was investigating cosmic particles with cloud chambers divided in two by a lead barrier. A cosmic particle crashing through the lead would lose a considerable portion of its energy, and it seemed to Anderson that the less energetic particle emerging from the barrier would curve more markedly in the presence of a magnetic field and, generally, reveal its properties more clearly. However, some cosmic particles in bursting through the lead smashed into atomic nuclei and sent out secondary radiations.

One of Anderson's photographs showed a particle of startling characteristics to have been ejected from the lead. From the extent of its curvature it seemed to have a mass equal to that of the electron, but it curved in the wrong direction. It was Dirac's positively-charged electron.

Anderson named it, naturally enough, the *positron*, and that is the name by which it is now universally known. The positron, since it is a particle opposed in certain key properties to that of a more familiar particle, belongs to a class now termed *anti-particles*. Were it discovered now it would be called the *anti-electron* and, indeed, it may be referred to in this fashion at times.

The question of symbolism is a little confused. One could use a full symbol, including the charge as subscript and mass as superscript, so that the electron is $_{-1}e^0$ while the positron is $_1e^0$. The disadvantage to this is that it is cumbersome. Most physicists do not feel that they need to be reminded of the size of the charge and the mass (particularly since the mass is not truly 0 but only very close to it). For that reason, it is very common, to symbolize the electron simply as e^- and the positron as e^+. This, too, has its difficulties, for, as it turned out later, there are anti-particles that have the same charge (or lack of charge) as the particles they oppose. For this reason, it is sometimes more convenient to indicate the anti-particle with a bar above the symbol. Thus, an electron would be e and a positron would be $\bar{e}$.

Positrons play a role in radioactivity, and this can best be understood if we once again go over the role played by the electron.

When the number of neutrons in a nuclide is too great for stability, the situation can be corrected by the conversion of a neutron to a proton with the emission of an electron. If we write the symbols in full (so that we can observe the manner in which mass and charge are conserved), we can say:

$$_0n^1 \longrightarrow {}_1p^1 + {}_{-1}e^0 \qquad \text{(Equation 13-1)}$$

In this process, the atomic number of the nuclide increases by 1 because an additional proton appears, but the mass number remains unchanged, since the proton appears at the expense of a disappearing neutron.

Consider phosphorus, for instance. Its only stable isotope is phosphorus-31 (15 protons, 16 neutrons). If we found ourselves with phosphorus-32 (15 protons, 17 neutrons) and observed it to be radioactive, we would expect that, because of its neutron excess, it would eliminate an electron in the form of a beta particle. Sure enough, it does. It emits a beta particle and becomes the stable isotope sulfur-32 (16 protons, 16 neutrons).

All naturally occurring radioactive isotopes, long-lived or short-lived, possess a neutron excess and, in the process of rearranging the nuclear contents to achieve stability, sooner or later emit electrons (though they may also emit alpha particles).

What if an artificial radioisotope is formed which has a neutron deficit? To achieve stability, a neutron must be gained and this must be at the expense of a proton. This can be done by a direct reversal of Equation 13-1; the absorption of an electron by a proton as in K-capture (see page 142).

$$_1p^1 + {}_{-1}e^0 \longrightarrow {}_0n^1 \qquad \text{(Equation 13-2)}$$

There is also the possibility, however, of another type of reversal. While a neutron can be converted to a proton with the emission of an electron, a proton can, analogously, be converted to a neutron with the emission of a positron:

$$_1p^1 \longrightarrow {}_0n^1 + {}_1e^0 \qquad \text{(Equation 13–3)}$$

The emission of a positron (or "positive beta particle") has the reverse effect of the emission of an electron. The atomic number of the nuclide is decreased by one, since a proton disappears. Again, the mass number is unchanged since a neutron appears in place of the proton.

As it happened, the very first artificial radioisotope formed, phosphorus-30, suffered from a neutron deficit. Where the stable phosphorus-31 is made up of 15 protons and 16 neutrons, phosphorus-30 is made up of 15 protons and only 15 neutrons. Phosphorus-30, with a half-life of 2.6 minutes, emits a positron and becomes the stable silicon-30 (14 protons, 16 neutrons). In forming phosphorus-30, the Joliot-Curies nearly anticipated Anderson in the discovery of the positron.

A large number of positron-emitters have been prepared among the radioactive isotopes artificially produced in the laboratory. Perhaps the best-known is carbon-11, which before the discovery of carbon-14 was much used as an isotopic tag.

The most important positron-producing process in nature is the hydrogen fusion that proceeds in the sun and in other stars. The overall change of four hydrogen-1 nuclei to a helium-4 nucleus is that of 4 protons to a 2-proton/2-neutron nucleus. Two of the protons, therefore, have been converted to neutrons with the emission of two positrons:

$$_1H^1 + {}_1H^1 + {}_1H^1 + {}_1H^1 \longrightarrow {}_2He^4 + {}_1e^0 + {}_1e^0$$
$$\text{(Equation 13–4)}$$

Matter Annihilation

The electron is a stable particle—that is, left to itself it undergoes no spontaneous change. This is in accord with the law of conservation of electric charge, which states that net charge can neither be created nor destroyed. The electron is the least massive particle known to carry a negative electron charge, and physicists work on the assumption that no smaller negatively-charged particle can conceivably exist. In breaking down, an electron would have to become a less massive particle, and then there is no room, so to speak, for an electric charge—so the electron doesn't break down.

This same argument holds for the positron, which cannot break down, for it is the least massive particle known to carry a positive electric charge, and has nowhere to dispose of it if it does break down. The positron is therefore also considered a stable particle and would, presumably, remain in existence forever if it were alone in the universe.

However, the positron is not alone in the universe. When formed, it exists in a universe in which electrons are present in overwhelming numbers. Under ordinary conditions on earth, it cannot move for more than a millionth of a second or so before it collides with an electron. What happens then?

If we consider a positron and electron together, the net electric charge is zero. The two can therefore merge and cancel each other's charge. In doing so, they apparently also cancel each other's mass in *mutual annihilation*. It is not true annihilation, however, for something is left since the law of conservation of mass-energy remains in force regardless of the situation with respect to electric charge. If the mass of the electron and positron disappear, an equivalent amount of energy must appear.

The total mass of an electron and a positron is 1.822×10^{-27} grams. Making use of Einstein's equation, $e = mc^2$ (see page II–111), we can determine that the energy equivalence of the two particles is 1.64×10^{-6} ergs, or 1.02 Mev.

In this conversion of mass to energy, however, there are other conservation laws that must be observed. The law of conservation of angular momentum (see page I–81) governs the distribution of spin, for instance.

The photon's spin is accepted, by definition, to be either $+1$ or -1. If an electron and positron can annihilate each other with the formation of a photon with an energy of 1.02 Mev (a gamma ray photon), and if it is assumed, as seems likely, that electron and positron have equal spin, then both must have a spin of $1/2$. If both have a spin of $+1/2$, then a photon of spin $+1$ is formed, and if both have a spin of $-1/2$, then a photon of spin -1 is formed.

However, the difficulty here is that the law of conservation of linear momentum (see page I–69) must also be respected. If the positron-electron system has a net momentum of zero with respect to its surroundings, then a single photon could not move after it was produced. Since a photon must move, and at the velocity of light, it follows that the production of a single photon is unlikely.

Instead three photons, each of 0.34 Mev energy (still gamma

rays), must be produced simultaneously, and shoot off toward the apices of an equilateral triangle. If the three photons have spins of $+1$, $+1$, and -1 respectively, the net spin is $+1$, while if the spins are -1, -1, and $+1$, the net spin is -1. In either case both angular momentum and linear momentum are conserved.

If the electron and positron spin in the same sense (that is, if both have a positive spin or both have a negative spin), then three photons can be produced, but not two. Two photons taken together can have a spin of 0 ($+1$ plus -1), $+2$ ($+1$ plus $+1$) or -2 (-1 plus -1), whereas the total spin of an electron and positron spinning in the same sense can only be $+1$ ($+1/2$ plus $+1/2$) or -1 ($-1/2$ plus $-1/2$). Angular momentum is not conserved.

On the other hand, if an electron and positron spin in opposite senses ($+1/2$ and $-1/2$), they can produce two photons ($+1$ and -1), for the net angular momentum is 0 both before and after; consequently, angular momentum is conserved. The two photons are gamma rays of 0.51 Mev each and dart off in opposite directions to conserve linear momentum.

I have gone into some detail here to show how nuclear physicists use the various conservation laws to decide what events on the subatomic scale can take place and what events cannot. They work on the assumption that any nuclear event that can happen will indeed happen if one waits long enough and looks hard enough. If, therefore, some particular event does not take place despite hard, long search, and nevertheless does not seem to be forbidden by any conservation law, a new conservation law is tentatively introduced. On the other hand, if an event that is forbidden by a conservation law takes place, it is necessary to recognize that this conservation law will hold only in certain circumstances and not in others, or that a deeper and more general conservation law must be sought.

It has been observed that when electrons and positrons annihilate each other, gamma rays of just the energies predicted by theory are formed. This is one of the neatest verifications of Einstein's special theory of relativity and of the mass-energy equivalence that is a part of it.

The reverse of all this is also to be expected. Energy ought to be converted into mass. No amount of energy can form an electron alone, or a positron alone, for in either case where is the electric charge to come from? Neither a net negative nor a net positive charge can be created.

However, an electron and a positron can be created simultaneously. The net charge of such an *electron-positron pair* is still zero. A gamma ray of at least 1.02 Mev energy is required for this, and if a more energetic gamma ray is used, the pair of particles possesses a kinetic energy equal to the energy excess over 1.02 Mev. The observed energy bookkeeping works out perfectly to the credit of Einstein.

Indeed, it is because of the superabundant energy of cosmic particles that energetic positrons are formed, and it is those which, detected by Anderson, marked the discovery of the first anti-particle.

When Dirac first worked out the theoretical reasoning that gave rise to the concept of the anti-particle, he felt that the electron's opposite number was the proton. However, this did not prove to be the case. The proton and the electron are exact opposites in electric charge, but in hardly anything else. The proton is, for instance, 1836 times as massive as the electron. (Why this should be so, and why the mass ratio should be 1836, no more and no less, is one of the more interesting unanswered questions of nuclear physics.)

The electron and the proton attract each other, as would any objects carrying opposing electric charges, but they do not and cannot annihilate each other. At best, the electron is captured by the proton and can approach to a minimum distance, representing the lowest possible energy-state. (If proton-electron annihilation were possible, matter could not exist.)

The electron and positron, which can annihilate each other, may also, at least temporarily, capture each other without annihilation. The "atom" consisting of an electron and positron circling each other (if we accept an ordinary particle view and ignore the wave manifestations) about a mutual center of gravity, is called *positronium*.

Two varieties exist, ortho-positronium, in which the two particles spin in the same sense, and para-positronium, in which they spin in opposite senses. The average existence of the former is a ten-millionth of a second (or a tenth of a microsecond) before annihilation takes place. The latter lasts only a ten-thousandth of a microsecond. The former forms three photons on annihilation, the latter forms two. The Austrian-American physicist Martin Deutsch (1917–   ) was able to detect positronium in 1951 by the light (if I may permit myself a play on words) of the gamma rays they emitted.

Anti-Baryons

There is nothing in Dirac's theory that cannot be applied to the proton as well as to the electron. If the electron has an antiparticle, then so must the proton. Such an *antiproton* could annihilate a proton and produce photons in pairs or in triplets, just as is true for the positron and electron.

However, since a proton has 1836 times the mass of an electron and the antiproton 1836 times the mass of a positron, the energy produced must be 1836 times that produced in electron/positron annihilation. The total energy produced is 1.02×1836, or 1872 Mev. This can also be expressed as 1.872 Bev. We are, as you see, in the billion electron-volt range.

In reverse, the formation of a proton/antiproton pair requires an input of 1.872 Bev at the very least. In fact, more energy is needed since the pair must be formed as the result of the collision of two highly energetic particles, and by using an excess of energy we would make the chances of antiproton production that much better. Physicists estimated that an energy of 6 Bev would turn the trick comfortably.

This energy is present in the more energetic cosmic particles. However, the more energetic cosmic particles are not common and to expect one of them to form a proton/antiproton pair just at the moment when someone is waiting with an appropriate detecting device is to ask a great deal of coincidence.

As it turned out, the discovery of the antiproton was not made until physicists had developed accelerators capable of producing particles in the billion electron-volt range. Particles in the Bev range could then be concentrated on some target at a time when specialized detecting setups were operating. At the University of California a proton synchroton, appropriately called the "Bevatron," was used for the purpose.

The energetic particles produced by the Bevatron were allowed to fall on a copper block, and a vast number of particles were formed by the colossally energetic collisions that resulted. It was then necessary to sort out any antiprotons that might have formed from among all the other explosion debris. The debris was led through a magnetic field that sorted out the negatively-charged particles. Among these, the antiproton was the most massive and traveled most slowly. The debris was therefore led across two scintillation counters some 40 feet apart, and only when those two counters registered with a time interval exactly equivalent to that of the time it would take an antiproton to cover that distance

(0.051 microseconds) was an antiproton considered to be detected.

This was accomplished in 1956 by Segrè (the discoverer of technetium, who by that time had emigrated to the United States) and the American physicist Owen Chamberlain (1920–   ).

The antiproton is, as might be expected, the twin of the proton, equal to it in mass but differing in charge. The proton is positively charged, but the antiproton is negatively charged. Proton and antiproton can therefore be symbolized as $_1p^1$ and $_{-1}p^1$ respectively, or as p^+ and p^-, or as p and $\bar{p}$.

The proton is a stable particle and, if left to itself, will presumably exist forever. There seems no obvious conservation law to account for this stability. Might not a proton break down to a positron of 0.51 Mev energy and yield up the remainder of its vast energy in the form of photons? Would not electric charge be conserved?

The fact that this has never been observed to happen means a new conservation law may well be involved. This is the *law of conservation of baryon number*, which states that in any subatomic event, the net number of baryons must be the same before and after. This has always been observed to be so in all the subatomic events studied, and physicists are convinced the law is valid.

If a proton breaks down to a positron, 1 baryon is changed to 0 baryons. This violates the law of conservation of baryon number, and so it doesn't happen. In fact, the proton is the least massive of all baryons, so it can't break down, and thus its stability is a reflection of a conservation law.

Similarly, an antiproton is stable and can't break down to an electron, for instance. It is an *anti-baryon*, the least massive of all anti-baryons, and the law of conservation of baryon number applies to anti-baryons as well.

In the actual universe, however, the antiproton encounters one of the protons (present in overwhelming numbers) almost at once, and mutual annihilation takes place. The net charge of a proton/antiproton pair is zero, so annihilation is possible without violation of the law of conservation of electric charge. In addition, an antiproton is considered as having a baryon number of −1, while a proton has a baryon number of 1. Consequently, the baryon number of a proton/antiproton pair is zero, and annihilation can take place without violating the conservation of baryon number.

The energy resulting from proton/antiproton annihilation may make itself evident in the formation of other particles, rather than as photons only. It sometimes happens, for instance, that where the proton and antiproton score a near miss, it is the charge only and not the mass that is annihilated. One might suppose that an uncharged particle is formed, but one such particle alone cannot be formed. The baryon number of a proton/antiproton pair is 0, but if a neutron, say, is formed, its baryon number is 1, and baryon number is not conserved. Instead, two particles must be formed, a neutron and an *antineutron*, with baryon numbers of 1 and —1, respectively, for a net baryon number of 0. In this way baryon number is conserved. This sort of "semi-annihilation" was first noted in 1956, shortly after the discovery of the antiproton, and that marked the discovery of the antineutron.

It is fair enough to ask what the difference between the neutron and antineutron might be. In the case of the other two particle/anti-particle pairs, the electric charge offers a handy means of differentiation. The electron is negative, the positron, positive. The proton is positive, the antiproton, negative.

There is, however, another difference as well, for these are all particles which possess spin. A spinning particle, if viewed as a tiny sphere, can be pictured as turning about an axis and possessing two poles. If viewed from above one pole, it would seem to be spinning counterclockwise; if viewed from above the other, it would seem to be spinning clockwise. Let us suppose the particle always to be pictured with the counterclockwise pole on top.

A spinning electric charge sets up a magnetic field with a north magnetic pole and a south magnetic pole. In the proton, viewed with the counterclockwise pole on top, the north magnetic pole is on top and the south magnetic pole is on the bottom. In the antiproton, on the other hand, with the counterclockwise pole still on top, it is the south magnetic pole that is on top and the north magnetic pole that is on bottom. In other words, if a particle and an anti-particle are so oriented as to spin in the same sense, the magnetic field of one is reversed with respect to the other. This is also true of the electron and positron.

Although the neutron has no electric charge, it does have a magnetic field associated with it. This is so because although the neutron has a net charge of zero, it apparently has local regions of charge associated with it. The American physicist Robert Hofstadter (1915–   ), in experiments from 1951 onward, has probed individual nucleons with beams of high-energy electrons. His results seem to indicate that both protons and neutrons are made up of shells of electric charge, and that they differ only in the net total charge.

Because of the neutron's magnetic field, it is possible to speak of both a neutron and an antineutron, the orientation of the magnetic field of one being opposed to that of the other. Because neither neutron nor antineutron has a charge, the symbol $_0n^1$ can apply equally well to both. The two are therefore invariably symbolized as n and $\bar{n}$ respectively.

The neutron decays, with a half-life of 1013 seconds, to a proton and an electron. A baryon is thus converted to a slightly less massive baryon, so baryon number is conserved. A net charge of 0 produces a net charge of 0, so electric charge is conserved. To be sure an electron is created but there is an added refinement to this reaction that will be discussed in the next chapter and that will bring the electron under the guardianship of a conservation law, too (see page 238).

In the same way, an antineutron can decay, with a half-life of 1013 seconds, to an antiproton and a positron, with conservation of baryon number (-1 before and after) and charge (0 before and after). The two events might be written:

$$n \longrightarrow p + e \qquad \text{(Equation 13-5)}$$
$$\text{and } \bar{n} \longrightarrow \bar{p} + \bar{e} \qquad \text{(Equation 13-6)}$$

Antimatter

We have now extended Table XIII by three more particles, the positron, the antiproton, and the antineutron, each the mirror-image, so to speak, of one of the particles in the table. Nor may we expect further mirror-images among these particles, for the photon and graviton cannot contribute anti-particles to the table. From theoretical considerations, each of these massless particles is considered to be its own anti-particle. The "anti-photon" and "anti-graviton" are, in others words, identical with the photon and graviton respectively.

We now have, then, four leptons (including one anti-lepton) and four baryons (including two anti-baryons).

Our universe (or at least that part of it which we can study) is very lopsided as far as particle/anti-particle distribution is concerned. It is composed almost entirely of particles, while anti-particles are extremely rare and, even when they are formed, live for only a fraction of a microsecond.

It is fair to wonder why this should be. Most physicists seem to assume that matter was created out of energy, either continuously—little by little—or all at once—at some long-past time.

We might assume, for instance, that matter is produced in the form of neutrons that then decay to form protons and electrons, and that the universe is built out of all three, plus additional energy in the form of photons and gravitons.

But if a neutron is formed, the law of conservation of baryon number would seem to require that an antineutron be simultaneously created. This antineutron would then break down to form antiprotons and positrons. The net result would be that particles and anti-particles would be formed in equal quantities, and any set of nuclear events that could be imagined to have led to the creation of the universe would yield the same result.

Still, if particles and anti-particles were created simultaneously, they would surely interact in mutual annihilation and return to the energy from which they sprang. Under these conditions, the universe could not be created.

It may be, therefore, that although particles and anti-particles were formed simultaneously, they were formed under such conditions that they separated at once, so that the chance for interaction was lost.

Thus, the effect of gravity on individual subatomic particles is so small that it has never been actually measured. It is possible that whereas particles are very feebly attracted to a gravitational field, anti-particles are very feebly repelled by it. In other words, anti-particles produce "anti-gravity." If particles and anti-particles are formed in vast numbers, the gravitational field of one may strongly repel the gravitational field of the other, so that in the end, two universes, driven violently apart, may be formed. The Austrian-American physicist Maurice Goldhaber (1911–   ) has speculated on just this possibility and refers to the two universes as a "cosmon" and an "anti-cosmon." We live in the cosmon, of course.

In the cosmon, atomic nuclei are made up of protons and neutrons and are surrounded by electrons. In the anti-cosmon, consisting as it does almost entirely of anti-particles, there would be nuclei made up of antiprotons and antineutrons surrounded by positrons. Such "anti-atoms" would make up what is called *antimatter*.

A universe of antimatter, totally unobservable by us perhaps, would be in all resects analogous to ours, consisting of "anti-galaxies" made up of "anti-stars" about which "anti-planets" circled, bearing, perhaps, "anti-life" and including even "anti-intelligent observers" studying their universe just as we study ours. They would note that their universe consisted almost entirely of what we consider anti-particles and that particles would have but a rare and fugitive existence. However, it is safe to bet that they would consider their universe to be made up of particles and matter and ours to be of anti-particles and antimatter—and they would be as justified in supposing so as we are.

Another alternative is to suppose that there is only one universe (ours) within which matter and antimatter are distributed equally but in separate chunks. The only safe way of separating these chunks is to suppose that individual galaxies (or galaxy-clusters) are made up of only one variety of substance, either matter or antimatter, but that both galaxies and anti-galaxies might exist in the universe.

If this is so, observation of the fact would be difficult. The only information we receive from other galaxies rests upon their gravitational influence and on the radiation they emit——that is, upon the gravitons and photons that flow from them to us. And, since gravitons and photons are considered to be their own anti-particles, they are produced with equal ease by both matter and antimatter. In other words, an anti-galaxy emits the same gravitons and photons that a galaxy does, and the two cannot be distinguished in that fashion. (Unless it turns out to be true that matter and antimatter repel each other gravitationally and that there is, after all, such a thing as the anti-graviton. The chances of this seem small.)

It is possible, of course, that a galaxy and an anti-galaxy may occasionally approach each other. If so, the mutual annihilation that results should emit energy of a magnitude much more intense than that produced under ordinary conditions. There are indeed galaxies that release unusually colossal energies, and every once in a while the possibility of antimatter raises its head among speculative scientists.

In 1962, certain unusual objects called "quasi-stellar objects," or *quasars*, were discovered. These radiate with the energy of a hundred ordinary galaxies, though they are only one to ten light years in diameter (as opposed to an ordinary galactic diameter of as much as 100,000 light years).

However, every effort is being made to explain this radiation by processes that do not involve antimatter. Antimatter will be turned to only as a last resort, since it would be so difficult to confirm such a speculation.

14

Other Particles

The Neutrino

In Chapter 11, disappearance in mass during the course of nuclear reactions was described as balanced by an appearance of energy in accordance with Einstein's equation, $e = mc^2$. This balance also held in the case of the total annihilation of a particle by its anti-particle, or the production of a particle/anti-particle pair from energy.

Nevertheless, although in almost all such cases the mass-energy equivalence was met exactly, there was one notable exception in connection with radioactive radiations.

Alpha radiation behaves in satisfactory fashion. When a parent nucleus breaks down spontaneously to yield a daughter nucleus and an alpha particle, the sum of the mass of the two products does not quite equal the mass of the original nucleus. This difference appears in the form of energy—specifically, as the kinetic energy of the speeding alpha particle. Since the same particles appear as products at every breakdown of a particular parent nucleus, the mass-difference should always be the same, and the kinetic energy of the alpha particles should also always be the same. In other words, the beam of alpha particles should be *monoenergetic*. This was, in essence, found to be the case.

In some instances, to be sure, the beam of alpha particles could be divided into two or more subgroups, each of which was monoenergetic, but with the energy of the subgroups differing among themselves. It was shown without much trouble that this was so because the parent nucleus could exist at various energy levels to begin with. An excited parent nucleus had a bit more energy content than a non-excited one, and the alpha particles produced by the former had correspondingly more kinetic energy. For each different energy level of the parent nucleus, there was a separate subgroup of monoenergetic alpha particles, but in each case, mass-energy equivalence (or, in a broader sense, the law of conservation of energy) was upheld.

It was to be expected that the same considerations would hold for a parent nucleus breaking down to a daughter nucleus and a beta particle. It would seem reasonable to suppose that the beta particles would form a monoenergetic beam, too, or, at worse, a small group of monoenergetic beams.

Instead, as early as 1900, Becquerel indicated that beta particles emerged with a wide spread of kinetic energies. By 1914, the work of James Chadwick demonstrated the "continuous beta particle spectrum" to be undeniable.

The kinetic energy calculated for a beta particle on the basis of mass loss turned out to be a maximum kinetic energy that very few attained. (None surpassed it, however; physicists were not faced with the awesome possibility of energy appearing out of nowhere.)

Most beta particles fell short of the expected kinetic energy by almost any amount up to the maximum. Some possessed virtually no kinetic energy at all. All told, a considerable portion of the energy that should have been present, wasn't present, and through the 1920's this missing energy could not be detected in any form.

Disappearing energy is as insupportable, really, as appearing energy, and though a number of physicists, including, notably, Niels Bohr, were ready to abandon the law of conservation of energy at the subatomic level, other physicists sought desperately for an alternative.

In 1931, an alternative was suggested by Wolfgang Pauli. He proposed that whenever a beta particle was produced, a second particle was also produced, and that the energy that was lacking in the beta particle was present in the second particle.

The situation demanded certain properties of this hypothetical particle. In the emission of beta particles, electric charge was conserved; that is, the net charge of the particles produced after emission was the same as that of the original particle. Pauli's postulated particle therefore had to be uncharged. This made additional sense since, had the particle possessed a charge, it would have produced ions as it sped along and would therefore have been detectable in a cloud chamber, for instance. As a matter of fact, it was not detectable.

In addition, the total energy of Pauli's projected particle was very small—only equal to the missing kinetic energy of the electron. The total energy of the particle had to include its mass, and the possession of so little energy must signify an exceedingly small mass. It quickly became apparent that the new particle had to have a mass of less than 1 percent of the electron and, in all likelihood, was altogether massless.

Enrico Fermi, who interested himself in Pauli's theory at once, thought of calling the new particle a "neutron," but Chadwick, at just about that time, discovered the massive, uncharged particle that came to be known by that name. Fermi therefore employed an Italian diminutive suffix and named the projected particle the *neutrino* ("little neutral one"), and it is by that name that it is known.

An uncharged, massless particle struck physicists as being a "ghost particle," since it could be detected by neither charge nor mass. Its existence would have been rather difficult to swallow, even for the sake of saving the law of conservation of energy, and the neutrino might have been ignored had it not turned out to save three other conservation laws as well.

This came up most clearly in the application of neutrino theory to the breakdown of neutrons. The neutron breaks down, with a half-life of 12 minutes, to a proton and an electron, and the electron can emerge with any of a wide range of kinetic energies. It should therefore follow from Pauli's theory that the neutron

Neutron breakdown

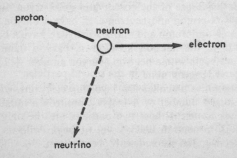

should break down to three particles—a proton, an electron, and a neutrino.

The difference in breaking down to three particles rather than two is significant in connection with the law of conservation of momentum (see page I–69). If a stationary neutron broke down to two particles only, these two would have to be ejected in opposite directions, their lines of travel forming a straight line. Only so could momentum be conserved.

If the same neutron broke down to three particles, then any two of those particles would have to be ejected to one side of an imaginary straight line, yielding a net momentum in a particular direction that would be exactly balanced by the momentum of the third particle shooting off in the opposite direction.

Studies of neutron breakdown indicated clearly enough that the proton and electron, when formed, went shooting off to one side of a straight line and that the existence of a third particle shooting off to the other side was absolutely necessary if momentum were to be conserved.

Once the matter of particle spin was understood, it became clear that the neutrino was useful in connection with the law of conservation of angular momentum (see page 81) as well. The neutron, proton and electron all have spins of either $+1/2$ or $-1/2$. Suppose a neutron broke down to a proton and electron only. The proton and electron together could have a spin of $+1$, 0, or -1 ($+1/2 + 1/2, +1/2 - 1/2,$ or $-1/2 - 1/2$). In no case could they match the neutron's original spin of $+1/2$ or $-1/2$, and angular momentum would not be conserved.

But suppose the neutrino also had a spin of either $+1/2$ or $-1/2$. In that case, the sum of the spin of all three particles could easily be $+1/2$ or $-1/2$. It could, for instance, be $+1/2 + 1/2 - 1/2$, and thus angular momentum would be conserved.

Finally, there is a more subtle conservation law. In the previous chapter, I made use of the conservation of baryon number (see page 229). A proton and neutron each have a baryon number of $+1$, and an antiproton and antineutron each have a baryon number of -1. In the neutron breakdown, baryon number is conserved, for we begin with a neutron (baryon number $+1$) and end with a proton (baryon number $+1$).

Can we recognize a similar law involving electrons, with an electron possessing a number of $+1$ and a positron a number of -1? The answer is: not if those two particles are the only ones considered. Thus, in neutron breakdown, we begin with no electrons (or positrons), and we end with one electron.

However, suppose we consider an *electron family* that includes not only electrons and positrons, but neutrinos, too. To make matters work out, it will be necessary to have not only a neutrino but an *antineutrino* as well. The difference between the neutrino and antineutrino would involve the direction of the magnetic field associated with the spinning particles, exactly as in the case of the neutron and antineutron (see page 231). The neutrino can be given an electron family number of $+1$ and the antineutrino an electron family number of -1.

With that in mind, let's consider neutron breakdown again. The neutron begins with an electron family number of 0, since it is itself not a member of the family. In breaking down, it produces a proton (electron family number 0) and an electron (electron family number $+1$). If we add to that, not a neutrino, but an antineutrino (electron family number -1), we have preserved the *law of conservation of electron family number*, which is 0 before and after breakdown.

The antineutrino saves the laws of conservation of energy, momentum, and angular momentum just as the neutrino would, and it enables us to retain the law of the conservation of electron family number as well. If we symbolize the neutrino as v (the Greek letter "nu") and the antineutrino as $\bar{v}$, we can write the equation for neutron breakdown as follows:

$$n^\circ \longrightarrow p^+ + e^- + \bar{v}^\circ \qquad \text{(Equation 14–1)}$$

On the other hand, in the conversion of a proton to a neutron with the ejection of a positron (see page 224), we have produced a particle with an electron family number of -1. To balance that we must add the production of a neutrino (electron family number $+1$). We can therefore write:

$$p^+ \longrightarrow n^\circ + e^+ + v^\circ \qquad \text{(Equation 14–2)}$$

Indeed, if we introduce neutrinos or antineutrinos into nuclear reactions, we can, whenever necessary, save the four conservation laws of energy, momentum, angular momentum, and electron family number. With this fourfold benefit, neutrinos and antineutrinos have to be accepted, whether they can be detected or not.

Neutrino Interactions

Despite the tightness of the reasoning from laws of conservation, physicists recognized that great satisfaction would come with the actual detection of the neutrino or antineutrino. To make detection possible, however, a neutrino or antineutrino must interact with some other particle in a recognizable manner.

Thus, a neutron changes to a proton, emitting an antineutrino in the process. Why cannot the reverse hold true and an antineutrino be absorbed by a proton to form a neutron? If so, this antineutrino absorption could leave a recognizable mark.

Unfortunately, the chance of such antineutrino absorption is vanishingly small. A neutron will break down to a proton with a half-life of 12 minutes. This means that in 12 minutes there is an even chance of a particular neutron producing an antineutrino. It follows that if an antineutrino remained in the immediate neighborhood of a proton for 12 minutes, there could be an even chance of the absorption taking place.

However, an antineutrino will not remain in the neighborhood of a proton for 12 minutes or, for that matter, even for a millionth of a second. Massless particles such as the neutrino, the antineutrino, the photon, or the graviton all begin moving at the speed of light at the moment of creation and keep moving at that speed until the moment of absorption. This means that an antineutrino remains in the immediate neighborhood of a proton only for about 10^{-23} seconds, and the chances of interaction in that short interval of time are exceedingly small. They are so small, in fact, that a neutrino or antineutrino would have to travel through some 3500 light-years of solid matter, on the average, before undergoing absorption.

The situation with regard to a photon is completely different. A photon also travels at the speed of light, but when the energy situation is such that a photon must be emitted by an atom, that photon is emitted in only about 10^{-8} seconds. Hence a photon need be in the vicinity of an atom for only about 10^{-8} seconds to stand a good chance of being absorbed. In addition, the photon has a considerably longer wavelength than the neutrino (viewing both as wave forms) and takes a longer time to pass an object, even though both are traveling at the same speed.

Gamma rays will, in fact, penetrate only ten feet of lead before being absorbed. Ordinary light, which is of much longer wavelength than gamma rays and takes even longer to pass a single atom, is even more readily absorbed and rarely penetrates more than a couple of dozen atom-thicknesses into a solid.

All this has an important consequence in astronomy. In the course of the fusion of hydrogen to helium, protons are converted to neutrons, so that neutrinos are formed as well as photons.

Photons carry off about 90 to 95 percent of the energy produced in the sun's core, while neutrinos carry off the remaining 5 to 10 percent.

The photons, once formed, are absorbed and re-emitted over and over again by the matter making up the sun; consequently, it takes something like a million years for the average photon to make its way from the core of the sun, where it is formed, to the surface, where it is radiated out into space. This insulating effect of solar material (thanks to the way in which photons so readily interact with matter) is dramatically demonstrated by the fact that the sun's core is at a temperature of 15,000,000° C, while the surface, only 430,000 miles away, is at a temperature of merely 6000° C.

The neutrinos formed in the core, however, are not re-absorbed by the matter of the sun. They shoot off instantly, at the speed of light, passing through the solar matter as though it were vacuum and taking less than three seconds to reach the solar surface and pass into space. This instant loss of energy has a small cooling effect on the sun's core, but not enough to matter.

A certain number of the solar neutrinos reach the earth and, after doing so, pass right through the planet in 1/125 of a second or less. About ten billion neutrinos pass through every square centimeter of the earth's cross section (passing through us, too). We are steadily bombarded both day and night, for the intervention of the bulk of the earth between ourselves and the sun does not interfere. However, the neutrinos pass through us without interacting, so they do not disturb us in any way.

It is possible that neutrinos and antineutrinos can be formed by methods that do not involve protons and neutrons. For instance, an electron-positron pair may be formed from gamma ray photons. The electron and positron may then react to form a neutrino and antineutrino:

$$e^- + e^+ \longrightarrow v^0 + \bar{v}^0 \qquad \text{(Equation 14–3)}$$

Energy, charge, momentum and angular momentum are all conserved in this reaction, and so is electron family number. The net electron family number of an electron and positron is zero, and that of a neutrino and antineutrino is also zero.

Such an electron-positron interaction is extremely unlikely even at the temperature of the sun's core, so that it makes no important contribution to the neutrino supply. In the course of a star's evolution, however, the core grows hotter and hotter, and as it does so the probability of conversion of photons to neutrinos via the electron-positron pair increases.

The American physicist Hong-Yee Chiu has calculated that when a temperature of 6,000,000,000° C is reached, the conversion of photons to neutrinos becomes so massive that the major portion of the energy formed in the core of such a star appears as neutrinos. These leave the core at once and withdraw so much energy that the core collapses and the star with it, resulting in a tremendous outburst of energy. This, it is suggested, is the cause of a supernova.

To say that a neutrino is extremely unlikely to interact with another particle is not the same, of course, as saying that it will never interact at all. If a neutrino must travel through an average of 3500 light-years of solid matter to be absorbed, that length of travel remains an average. Some neutrinos may survive for much longer distances, but some may be absorbed long before traversing such a path. There is a finite chance, exceedingly small but not zero, of a neutrino interacting after traveling only a mile or even only a foot.

Evidence for such interactions was sought by the American physicists Clyde L. Cowan, Jr. (1919–   ) and Frederick Reines (1918–   ) in experiments beginning in 1953. As the proton target they used large tanks of water (rich in hydrogen atoms and, therefore, in nuclei consisting of single protons), and they placed these in the path of a stream of antineutrinos originating in a fission reactor. (These antineutrinos arise in the course of the rapid conversion of neutrons to protons within the nuclei of fission products.)

If an antineutrino were to join a proton to form a neutron, in the reverse of the interaction of Equation 14–1, an electron would have to be absorbed simultaneously. The necessity of such a double joining makes the reaction less likely than ever. However, the absorption of an electron is equivalent to the emission of a positron, so the expected neutrino interaction with a proton may be described as follows:

$$\bar{v}^\circ + p^+ \longrightarrow e^+ + n^\circ \qquad \text{(Equation 14–4)}$$

In such a reaction, baryon number is conserved, for a proton ($+1$) is replaced by a neutron ($+1$). Electron family number is also conserved, for an antineutrino (-1) is replaced by a positron (-1).

Cowan and Reines calculated that in the water targets they

were using, such an antineutrino/proton interaction ought to take place three times an hour. The trouble was that a large number of other events were also taking place, events originating in cosmic radiation, stray radioactive radiations, and so on. At first these unwanted events took place with many times the frequency of the antineutrino reactions being searched for. With time, this interference was reduced to manageable levels by using heavy shielding that excluded most extraneous subatomic particles and photons but offered no barrier whatever, of course, to antineutrinos.

It remained to identify the antineutrino interaction precisely and certainly. The interaction produces a positron and a neutron. The positron interacts almost at once with an electron, producing two gamma rays of a known energy content, coming off in opposite directions.

The neutron produced by the interaction lasts a few millionths of a second longer before it is absorbed by a cadmium atom (introduced into the water tank in the form of a solution of cadmium chloride precisely for the purpose of absorbing neutrons). The cadmium atom, excited by neutron absorption, releases a gamma ray (or possibly three) of known frequency. It is this combination of events, a double gamma ray of fixed frequency, followed after a fixed interval by a third gamma ray of fixed frequency, that is the identifying mark of the antineutrino. No other particle could produce just the duplicate of these results, at least as far as is known.

In 1956, the antineutrino was finally detected by this characteristic pattern of gamma radiation, and Pauli's original suggestion of a quarter of a century earlier was vindicated.

The Muon

While Pauli was advancing his solution for the problem of the continuous beta particle spectrum, a second problem, just as puzzling, had arisen.

The atomic nucleus contains protons held together in a volume of something like 10^{-40} cubic centimeters. The electromagnetic force of repulsion between such protons, jammed so closely together, is tremendous. As long as it was believed that electrons also existed within the nucleus, it could be supposed that the attraction between protons and electrons (also jammed tightly together) could make up for the inter-proton repulsion. The electrons, then, would serve as a "nuclear cement," and electromagnetic forces would explain the situation within the nucleus as they did the situation between atoms and molecules.

When, however, it became quite clear, in 1932, that the atomic nucleus was made up of protons and neutrons and that electrons did not exist there, the problem was thrown wide open. Since only electromagnetic repulsion could exist within the nucleus, why did not all atomic nuclei explode at once?

The only way of explaining the stability of the nucleus was to suppose that there was a nuclear force of attraction between nucleons, one that was in evidence only at extremely small distances and that was then much stronger than the electromagnetic force.

In the early 1930's, quantum mechanical analysis made it seem that a force, which seemed to act at a distance as the electromagnetic force did, actually acted through the emission and absorption of photons. Electrically-charged particles, exchanging photons, experienced *exchange forces,** a term introduced by Heisenberg in 1932. By analogy, it was decided that the gravitational force had to make itself evident through the emission and absorption of gravitons (see page 221).

Both the electromagnetic force and the gravitational force are long-distance forces, diminishing only as the square of the distance between the objects exerting the force, and making themselves evident even over astronomical distance.

The hypothesized nuclear force, however, had to be extremely short-range and, however strong within the nucleus, it had to be imperceptible outside the nucleus. In fact, in the larger nuclei, the nuclear force must barely reach across the full diameter, and it may be for this reason that nuclear fission takes place as easily as it does in the more massive atoms.

The Japanese physicist Hideki Yukawa (1907–   ) set himself the task of working out the mechanism of such an unusually strong and unusually short-range force. Without going into the quantum mechanics of the theory, we can present a simplified picture of the reasoning involved.

The principle of uncertainty states that position and momentum cannot be simultaneously determined with complete accuracy. The uncertainty in the determination of one multiplied by the uncertainty in the determination of the other is approximately equal to Planck's constant. It can be shown that time and

---

* Actually, the word "force" is going out of fashion among physicists. In subatomic physics, particularly, *interaction* is preferred to "force" when describing the consequences of emission and absorption of particles.

energy can be substituted in place of position and momentum. This means that the precise energy content of a system cannot be determined at an exact moment of time. There is always a small time interval during which the energy content is uncertain. The uncertainty in energy content multiplied by the uncertainty in time is again approximately equal to Planck's constant.

During the interval of time in which energy content is uncertain, a proton might, for instance, emit a small particle. It doesn't really have the energy to do this, but for the short instant of time during which its energy cannot be exactly determined, it can violate the law of conservation of energy with impunity—because, so to speak, no one can get there fast enough to enforce it.

By the end of the time period that particle emitted by the proton must be back where it started, and the proton must be obeying energy conservation. The particle, which is emitted and re-absorbed too quickly to be detected, is a *virtual particle*. Reasoning shows it can exist, but no system of measurement can detect it.

During the period of existence of the virtual particle it can move away from the parent proton, but it can only move a limited distance because it must be back when the time-uncertainty period is over. The more massive the particle (and the greater its energy content), the greater the uncertainty represented by this energy and the shorter the time interval permitted its existence, for the two together must yield the same product under all circumstances so that as one uncertainty goes up the other goes down in precise step.

Even if the virtual particle were traveling at the speed of light, it could not move very far from its proton, for Planck's constant is a very small quantity, and the time interval permitted the particle's existence is excessively tiny. Ordinarily, the virtual particle never reaches far enough from the proton to impinge on any other particle. The only exception arises when protons and neutrons are in the close proximity found within the atomic nucleus. Then, one of the particles leaving the proton may be picked up by a neutron before it has a chance to return to the proton. It is this emission and absorption of virtual particles that produces the nuclear force.

In 1935, Yukawa advanced his views that this virtual particle served as the exchange particle of the nuclear force. Unlike the exchange particles of the electromagnetic and gravitational forces, the exchange particle of the nuclear force had to have mass so that its permitted time of existence would be brief enough to make it sufficiently short-range. Yukawa showed that the particle would have to be about 270 times as massive as an electron in order for its permitted time of existence to be short enough to make it as short-range as observation showed the nuclear force must be.

Because such a particle is intermediate in mass between the light electrons and the massive particles of the nucleus, it came to be called a "mesotron," from a Greek word meaning "intermediate," and this was quickly shortened to *meson*.

Yukawa's theory indicated that in the process of exchange, a proton would become a neutron and a neutron would become a proton. In other words the meson, in being emitted by one and absorbed by the other, would have to carry the charge with it. You would expect a positive meson, therefore. In the case of antiprotons and antineutrons you would expect a negative meson as an anti-particle, holding the nucleus of antimatter together.

Then, too, it turned out that exchange forces existed between proton and proton and between neutron and neutron; for this a neutral meson was needed. This neutral meson served as its own anti-particle and served equally well to bind antiproton and antiproton or antineutron and antineutron.

The proton-neutron exchange force is somewhat stronger than the proton-proton exchange force, which means that the *p-n* combination within a nucleus has a lower packing fraction than the *p-p* combination. It therefore takes an energy input to convert a *p-n* to a *p-p* within a nucleus.

The conversion of *n* to *p* yields a small amount of energy (which is why a neutron breaks down spontaneously), but the quantity of energy so obtained is not always sufficiently great to change the *p-n* combination to a *p-p* combination. It is for this reason that in some nuclei the neutron does not change to a proton but stays put; and thus stable nuclei exist.

To check Yukawa's theory, the mesons would actually have to be detected. Within the nucleus, where they are virtual particles only, this cannot be done. However, if enough energy is added to the nucleus, the meson can be formed without violating conservation of energy. It then becomes a real particle and can leave the nucleus.

In 1936, Carl Anderson, who had earlier discovered the positron among the tracks produced by cosmic particles, now found a track which curved less sharply than an electron and more sharply than a proton. It was obviously produced by a particle of intermediate mass, and physicists assumed at first that it was Yukawa's predicted particle.

This proved not to be the case. Anderson's particle was only 207 times the mass of an electron, distinctly less than Yukawa's prediction. It came only in a positive and negative variety, without any sign of a neutral variety; and it was the negative variety, rather than the positive variety, that was the particle. Worst of all, it did not seem to interact with protons or neutrons. If it was Yukawa's exchange particle it should have been absorbed by any nucleon it encountered. Anderson's meson, however, passed through matter almost undisturbed.

It eventually turned out that there were not one but a number of different mesons and that Anderson's meson was not Yukawa's exchange particle. The different types of mesons were given different prefixes (often one or another of the Greek letters) and Anderson's was named a *mu-meson*, a term which is now commonly shortened to *muon*.

As the properties of the muon were studied more and more closely, it turned out that the muon seemed more and more similar to an electron. It was identical in charge, with the negative variety serving as the particle; the positive, as the anti-particle. The muon was the same as the electron in spin and in magnetic properties—in everything but mass and stability.

Indeed, for every interaction involving the electron, there is an analogous interaction involving the muon. The muon, while it lives, can even replace the electron within atoms to form a *mesonic atom*. Angular momentum must be conserved in the process. If we view the electron in the old-fashioned way as a particle circling the nucleus, the muon (moving at the same speed) must circle in an orbit closer to the nucleus. Its greater mass is thus countered by the shortened radius of revolution to keep the angular momentum the same (see page I–82).

Since the muon is 207 times the mass of the electron, it must be at only 1/207 the distance from the nucleus. In very massive atoms, it means that the orbit of the innermost meson must actually be within the nucleus! The fact that it can circle freely within the nucleus shows how small a tendency it has to interact with protons and neutrons.

The difference in mesonic energy levels in such atoms is correspondingly larger than in the electronic energy levels in ordinary atoms. Mesonic atoms emit and absorb X ray photons in place of the visible light photons emitted and absorbed by ordinary atoms.

To be sure, the muon is unstable, decaying in about 2.2 microseconds, and changing to an electron. However, on the subatomic scale 2.2 microseconds is quite a long time, and the muon does not seem too different in this respect from the completely stable electron.

It seems now that the muon is virtually a "heavy electron" and nothing more. But why there should be a heavy electron at all, and why it should be so much heavier, is as yet not known.

The Pion

Though the muon failed to fill the role of Yukawa's particle, there was success elsewhere. In 1947, the English physicist Cecil Frank Powell (1903–   ) duplicated Anderson's feat and uncovered meson tracks in photographic plates exposed to cosmic radiation in the Bolivian Andes. These new mesons were distinctly more massive than the muon. Their mass, in fact, was equal to 273 electrons, almost exactly the predicted mass for Yukawa's exchange particles.

On investigation they proved to interact strongly with nuclei, as Yukawa's exchange particle would be expected to do. The new meson was a particle when positively charged and an anti-particle when negatively charged, as was to be expected. Eventually, a neutral version of this meson was also found, one that was somewhat lighter than the charged varieties (only 264 times the mass of the electron).

The new meson was named the *pi-meson*, or, as it is now commonly known, the *pion*, and it is the pion that is Yukawa's exchange particle. Both the neutron and proton are now viewed as consisting, essentially, of clouds of pions. This was demonstrated in the 1950's by Robert Hofstadter, who bombarded protons and neutrons with electrons of 600 Mev energy, produced in a linear accelerator. These electrons, in being scattered, actually penetrated the proton passed through the outer portion of the pion cloud.*

Pions are unusual in the nature of their spin. Most of the

* Findings such as this raise the question of just which subatomic objects are *elementary particles*—that is, which are not composed of still smaller and simpler components. For that matter, do elementary particles exist at all? Does the phrase have meaning? Physicists have no good answer to this at the moment.

particles so far discussed—the neutrino, electron, muon, proton and neutron, together with their anti-particles—have spins of 1/2. Particles with such nonintegral spins behave according to *Fermi-Dirac statistics* (a mathematical analysis worked out by Fermi and Dirac) and are in consequence all lumped together as *fermions*. An outstanding property possessed by fermions generally is that of adhering to Pauli's exclusion principle (see page 80).

The photon has a spin of 1 and the graviton a spin of 2. These particles, and all others possessing integral spin, including a number of atomic nuclei, behave according to *Bose-Einstein statistics*, worked out by Einstein and by the Indian physicist Satyenda Nath Bose (1904–   ). Such particles are *bosons*, and the exclusion principle does not hold for them.

The pions were the first individual particles to be found with a spin of 0, and the first particles, possessing mass, which were bosons.

The ready reaction of a pion with nuclear particles is an example of what is known as the *strong interaction*. This is characterized by extreme rapidity. A pion traveling at almost the speed of light remains within short range of a proton or neutron for only 10^{-23} seconds, yet this is enough time for the strong interaction to take place. It is this strong interaction which is the nuclear force that holds the nucleus together against the repulsion of the electromagnetic interaction.

There are, however, interactions involving subatomic particles that take place in much longer intervals of time—only after a hundred-millionth of a second or more. These are the *weak interactions* which are very short-range, like the strong interactions but are only a trillionth as intense as the strong interactions. The weak interaction is, in fact, only a ten-billionth as intense as the electromagnetic interaction, but it is still tremendously stronger than gravitation, which retains its status as the weakest force in nature.

The pions are the exchange particles of the strong interactions and there should also be exchange particles for the weak interactions. This "weak exchange particle," symbolized by *w*, should be more elusive than either the pion or the photon, though not as elusive as the graviton. It should be one of the boson family of particles and should be more massive than such bosons as the photons, though less massive than such bosons as the pions. It is, for that reason, sometimes referred to as the *intermediate boson*. Some recent reports have indicated its detection, but this is not yet certain.

The proton, antiproton, positive pion and negative pion can all be involved in any of the four types of interaction: strong, weak, electromagnetic, and gravitational. The neutron, antineutron, and neutral pion, being uncharged, cannot be involved in electromagnetic interactions, but can engage in any of the remaining three. The electron, positron, positive muon and negative muon cannot take part in strong interactions, but can be involved in the remaining three.

The neutrino and antineutrino are the most limited in this respect. They do not take part in the strong interaction. Being uncharged, they do not take part in the electromagnetic interaction, and being massless, they do not take part in the gravitational interaction. The neutrino and antineutrino take part only in weak interactions and nothing more. The appearance of a neutrino or antineutrino in the course of the breakdown of a particle is thus a surefire indication that this breakdown is an example of a weak interaction. The breakdown of a neutron is, for instance, a weak interaction.

Once negative pions and positive pions are formed in the free state, they too break down in a weak interaction, with a half-life of 25 billionths of a second. They break down to muons and neutrinos, and if we allow pions to be represented as π (the Greek letter "pi") and muons as μ (the Greek letter "mu"), we can present the breakdown as follows:

$$\pi^+ \longrightarrow \mu^+ + \nu^0 \qquad \text{(Equation 14–5)}$$
$$\pi^- \longrightarrow \mu^- + \bar{\nu}^0 \qquad \text{(Equation 14–6)}$$

At first, physicists suspected that the neutrino produced in the course of pion breakdown might be distinctly more massive than the ordinary neutrino and, in fact, have a mass perhaps 100 times that of the electron. For a while, they called this particle the "neutretto." However, further study scaled the apparent mass downward, until finally it was decided that the small neutral product of pion breakdown was a massless neutrino.

If the muon is considered as only a "heavy electron," it might seem reasonable to include the muons in the electron family and give the negative muon (like the electron) an electron family number of $+1$, and the positive muon (like the positron) an electron family number of -1.

If so, then in Equation 14–5 the production of a positive muon (-1) and a neutrino $(+1)$ gives a net electron family number of 0, matching that of the original pion (which, not being a member of the electron family at all, has an electron family

number of 0). In the same way, in Equation 14–6, the production
of a negative muon $(+1)$ necessitates the simultaneous production of an antineutrino (-1) for, again, a net electron family
number of 0.

So far, so good, but difficulty arises when muon breakdown
is considered. A muon breaks down to form an electron and two
neutrinos. If electron family number is to be conserved, then one
of the neutrinos must be an antineutrino. The reaction (in the
case of a negative muon) can be written as follows:

$$\mu^- \longrightarrow e^- + \nu^0 + \bar{\nu}^0 \qquad \text{(Equation 14–7)}$$

Beginning with an electron family number of $+1$ for the
negative muon, one ends with a total electron family number
of $+1$ by taking the sum of $+1$, $+1$ and -1 for the electron, neutrino, and antineutrino respectively. The conservation
holds.

However, if this is so, should not the neutrino and antineutrino at least sometimes annihilate each other in a burst of
energy, as any particle/anti-particle combination would? Should
not then a negative muon break down to form an electron only,
at least sometimes, with the remaining mass of the muon appearing
as photons?

This is never observed and the suspicion therefore arose that
the neutrino and antineutrino produced by muon breakdown were
not true opposites. Could it be that the neutrino was produced in
association with the muon and that the antineutrino was produced
in association with the electron, for instance, and that muons and
electrons produced different kinds of neutrinos?

In 1962, this possibility was tested as follows: Very high
energy protons were smashed into beryllium atoms in such a way
as to produce an intense stream of pions. The pions broke down
rapidly to muons and neutrinos, and all then smashed into a wall
of armor plate about 13.5 meters thick. All particles but the
neutrinos were stopped. The neutrinos passed through easily, and
inside a detecting device would every once in a while interact
with a neutron to form a proton plus either a negative muon or an
electron.

If there were only one kind of neutrino then it ought to produce negative muons and electrons without discriminating between
the two:

$$\nu^0 + n^0 \longrightarrow p^+ + e^- \qquad \text{(Equation 14–8)}$$

$$\text{or} \quad \nu^0 + n^0 \longrightarrow p^+ + \mu^- \qquad \text{(Equation 14–9)}$$

As you see, electric charge and baryon number is conserved in either case. Electron family number would seem to be conserved in either case, too, for you would begin with a neutrino (electron family number + 1) and end with either an electron or a negative muon, each of which has an electron family number of + 1. In subatomic interactions, anything which can happen does happen, so physicists were sure that if there was only one kind of neutrino then negative muons and electrons would be produced in equal numbers.

They weren't! Only negative muons were produced.

This meant that when pions broke down to form muons and neutrinos, the neutrinos were *muon-neutrinos*, a special variety that could interact only to form muons, never electrons. Similarly, the ordinary neutrinos formed in association with electrons and positrons were *electron-neutrinos*, which could interact only to form electrons or positrons, never muons.

If we symbolize the muon-neutrino as v_μ and the electron-neutrino as v_e, we can rewrite Equations 14–1, 14–2, 14–3, 14–4, 14–5, and 14–6 as follows:

$$n^\circ \longrightarrow p^+ + e^- + \bar{v}_e{}^\circ \qquad \text{(Equation 14–10)}$$

$$p^+ \longrightarrow n^\circ + e^+ + v_e{}^\circ \qquad \text{(Equation 14–11)}$$

$$e^- + e^+ \longrightarrow v_e{}^\circ + \bar{v}_e{}^\circ \qquad \text{(Equation 14–12)}$$

$$\bar{v}_e{}^\circ + p^+ \longrightarrow e^+ + n^\circ \qquad \text{(Equation 14–13)}$$

$$\pi^+ \longrightarrow \mu^+ + v_\mu{}^\circ \qquad \text{(Equation 14–14)}$$

$$\pi^- \longrightarrow \mu^- + \bar{v}_\mu{}^\circ \qquad \text{(Equation 14–15)}$$

Equations 14–10, 14–11, 14–12, and 14–13, still exhibit conservation of electron family number. Equations 14–14 and 14–15 now exhibit a *conservation of muon family number*, where the members of the muon family include the negative muon and the muon-neutrino (each with a muon family number of + 1) and the positive muon and the muon-antineutrino (each with a family number of − 1). In Equations 14–14 and 14–15, as you see, there is a net muon family number of 0 both before and after the breakdown of the pion.

The one interaction which deals with both electrons and muons is Equation 14–7. This can now be rewritten:

$$\mu^- \longrightarrow e^- + v_\mu{}^\circ + \bar{v}_e{}^\circ \qquad \text{(Equation 14–16)}$$

This exhibits conservation of muon family number because you begin with a muon family number of + 1 (the negative muon) and end with a muon family number of $+1$ (the muon-neutrino). It also exhibits conservation of electron family number because you begin with an electron family number of 0 (no members of the electron family present) and end with an electron family number of $+1$ for the electron and -1 for the electron-antineutrino, making a net electron family number of 0 again.

By the same reasoning, the breakdown of the positive muon would be:

$$\mu^+ \longrightarrow e^+ + \bar{\nu}_\mu{}^0 + \nu_e{}^0 \qquad \text{(Equation 14–17)}$$

producing a positron, an electron-neutrino, and a muon-antineutrino.

In the breakdown of either the negative muon or the positive muon, you could expect no mutual annihilation among the neutrinos and antineutrinos since they are not true anti-particles. Mutual annihilation would violate conservation of both electron family number and muon family number.

However, whatever difference there may be between electron-neutrinos and muon-neutrinos, when both are massless, chargeless particles with a spin of $1/2$ remain a mystery.

The Frontier

Since 1947, a variety of other particles have turned up. With the exception of the muon-neutrino (which was more of a realization than a discovery), all the new particles are quite unstable, quite massive, and subject to strong interactions.

There are, for instance, a group of *K-mesons*, or *kaons*, which are 966.5 electrons in mass and, in this respect, lie roughly midway between protons and pions. Like the pions, the kaons have 0 spin and are bosons. Like the pions, also, there is a positive kaon, which is the particle, and the negative kaon, which is the anti-particle. There is also a neutral kaon, a trifle less massive than the charged ones and somewhat more unstable. However, the neutral kaon is not its own anti-particle as the neutral pion is. There is a neutral kaon and a neutral anti-kaon.

In addition to these, particles more massive than protons or neutrons were discovered. These fell into three groups, distinguished by Greek letters: *lambda particles, sigma particles* and *xi particles*.

There is one lambda particle (neutral), three sigma particles (positive, negative, and neutral), and two xi particles (negative and neutral). Each has its anti-particle. The lambda particle has a mass of 2182 electrons (or 1.18 protons). The three sigma particles are more massive, about 1.27 protons, while the xi particles are still more massive, about 1.40 protons. All of these are lumped together as *hyperons* (from a Greek word meaning "beyond," because they are beyond the proton in mass), and all are fermions.

Just as muons can replace electrons in atomic structure to form mesonic atoms, the lambda particle can replace a particle in the atomic nucleus to produce a short-lived *hypernucleus*.

The 1960's saw the discovery of numerous extremely short-lived particles, with lifetimes as short as 10^{-23} seconds. These are *resonance particles*. It is not certain that these are truly single particles. They may be merely momentary associations of two or more particles.

The proliferation of particles has been an embarrassment to physicists, for it is difficult to reduce them to order. New rules of behavior had to be deduced for them.

For instance, although the hyperons are produced under conditions that make it clear that they are strongly-interacting particles, and though they can break down to strongly-interacting products, they neverthelesss take a strangely long time about it. The lambda particle breakdown can be represented thus, for instance:

$$\Lambda^0 \longrightarrow p^+ + \pi^- \qquad \text{(Equation 14–18)}$$

where Λ (the Greek capital letter "lambda") represents the lambda particle. All the usual conservation laws are obeyed in this interaction. Since the pion has zero spin, for instance, angular momentum is conserved. Since hyperons are considered baryons, baryon number is conserved. (The pion is not a member of any of the family groups that are conserved, and can appear and disappear freely as long as other conservation laws are observed.)

Since Equation 14–18 seems to represent a strong interaction, it should take place in not much longer than 10^{-23} seconds or so; instead, however, it takes place in 2.5×10^{-10} seconds. This is ten trillion times as long as it ought to take—immensely long on the subatomic time scale.

In 1953, an explanation was independently advanced by the American physicist Murray Gell-Mann (1929–   ) and the Japanese physicist Kazuhiko Nishijima (1926–   ). They suggested a new quantity that had to be conserved, one which Gell-Mann called *strangeness*.

The various members of the electron family and muon family, as well as the pions, nucleons and their anti-particles, all have a strangeness number of zero. Other particles, with strangeness numbers other than zero, are lumped together as *strange particles*. The kaon has a strangeness number of $+1$; the lambda particle and sigma particle, one of -1; and the xi particle, one of -2. For the anti-particles the sign of the strangeness number is reversed, of course.

These numbers are not assigned arbitrarily but are deduced from experiment. A particle with strangeness number of $+1$ is never formed without the simultaneous production of particle with strangeness number -1, if the strangeness number was 0 to begin with. Strangeness, if the given strangeness numbers are used, is conserved.

When the lambda particle decays, as in .Equation 14–18, a lambda particle (strangeness number -1) becomes a proton and a negative pion (both with strangeness number 0). Strangeness is not conserved in that reaction and the breakdown ought not to occur.

However, the *law of conservation of strangeness* holds only for the strong interactions. Therefore, the breakdown can occur, provided it occurs by way of a weak interaction which takes a much longer time. Despite appearances, then, Equation 14–18 represents weak interactions and that accounts for the long lifetime of the lambda particle.

An older conservation law that was also found to have its limitations was the *law of conservation of parity*.

Parity is a quantity that is conserved in the same manner that "even" and "odd" are conserved in the realm of numbers. If an even number such as 8 is broken down into the sum of two smaller numbers such as $6 + 2$ or $5 + 3$, the two smaller numbers are either both even or both odd. If an odd number is so treated, as $7 = 4 + 3$, the smaller numbers are always one odd and one even. More complicated transformations will also yield universal rules of this sort.

In 1956, however, it turned out that some kaons decayed to two pions, and some to three pions. Since the pion is assigned odd parity, two pions together are even, while three pions together are odd. This meant that there were even-parity kaons and odd-parity kaons, and the two were given different names.

Yet the two types of kaons were absolutely identical in all properties but the manner of their breakdown. Was it necessary that parity be conserved? Tackling this problem, two Chinese-American physicists, Tsung-Dao Lee (1926–   ) and Chen Ning Yang (1922–   ), produced theoretical reasons why parity need not be conserved in weak interactions, even though it was conserved in strong interactions.

There was a possibility of testing this. As long ago as 1927, Eugene Wigner had considered the problem and had shown that the conservation of parity meant a lack of distinction between left and right, or (which is equivalent) between a situation and its mirror-image. This could only be true if all interactions took place symmetrically in space. If electrons were given off by nuclei, for instance, they would have to be given off in all directions equally so that the mirror-image would show no clear distinction from the actual state of affairs. If electrons were given off in one direction predominantly (say to the left), then the mirror-image would show them given off predominantly to the right, and the two states of affairs could be distinguished.

The Chinese-American physicist Madame Chien-Shiung Wu (1913–) tested the Lee-Yang theory, making use of cobalt-60, which gives off electrons in a weak interaction. She cooled the cobalt-60 to nearly absolute zero and subjected it to a magnetic field that lined up all the nuclei with their north magnetic poles in one direction and with their south magnetic poles in the other. At nearly absolute zero, they lacked the energy to move out of alignment.

It turned out then that electrons were not given off equally in all directions at all. They emerged in uniformly greater numbers from the south magnetic pole than from the north. This situation could be distinguished from its mirror-image, and the law of conservation of parity turned out not to hold for weak interactions.

Consequently, it is perfectly possible for a kaon to be odd parity sometimes and even parity other times, when only weak interactions are involved.

An attempt has been made to produce a more general conservation law by combining parity with *charge conjugation*, a quantity which involves the interchange of particles and anti-particles. This combination, usually abbreviated *PC conservation*, means that a shift in parity implies an appropriate change in connection with anti-particles. Thus the antimatter version of cobalt-60 would give off positrons predominantly from the north magnetic pole. Parity and charge conjugation are separately conserved in strong interactions and jointly conserved in weak interactions.

Gell-Mann went on, in 1961 (as did, independently, the Israeli physicist Yuval Ne'eman) to attempt to produce order among the growing dozens of strongly interacting particles by arranging them in a certain manner making use of eight properties conserved in strong interactions. Gell-Mann, making use of an advanced branch of mathematics called "group theory" to guide his arrangements, called the result the *eightfold way*.

In one case, for instance, Gell-Mann dealt with a group of four *delta particles*, new hyperons that came in varieties charged — 1, 0, + 1 and + 2. Above these he could place the somewhat more massive sigma particles (with charges — 1, 0, and + 1), and above these the still more massive xi particles (with charges — 1 and 0).

The regularity could be continued if at the apex of the triangle he was forming he could put an even more massive single particle with a charge of — 1. Since this was the end of the triangle, Gell-Mann named it the *omega-minus particle*, "omega" being the last letter of the Greek alphabet. From the arrangement of the various conserved properties, the omega-minus particle would have to possess particular values for these, including, most unusually, an unprecedented strangeness number of — 3.

In 1964, the omega-minus particle was detected and shown to have all the predicted properties with amazing fidelity, down to the strangeness number of — 3. It was a discovery as significant, perhaps, as those of Mendeleev's missing elements.

Here is the present frontier of physics—the world of subatomic particles which in the last two decades has become a jungle of strange and mystifying events but which, if the proper key is found, may yield a new, subtle, and intensely bright illumination of the physical universe.

Suggested Further Reading

Feynman, Richard P.; Leighton, Robert B.; and Sands, Matthew, *The Feynman Lectures on Physics* (Volume III), Addison-Wesley Publishing Co., Inc., Reading, Mass. (1963).

Ford, Kenneth W., *The World of Elementary Particles,* Blaisdell Publishing Co., Inc., New York (1963).

Friedlander, Gerhart; Kennedy, Joseph W.; and Miller, Julian Malcolm, *Nuclear and Radiochemistry,* John Wiley & Sons, Inc., New York (1964).

Glasstone, Samuel, *Sourcebook on Atomic Energy,* D. Van Nostrand Co., Inc., Princeton, New Jersey (1958).

Kaplan, Irving, *Nuclear Physics,* Addison-Wesley Publishing Co., Inc., Reading, Mass. (1963).

Lengyel, Bela A., *Lasers,* John Wiley & Sons, Inc., New York (1962).

INDEX

Television, 48-50
Teller, Edward, 196, 208
Tellurium, 15
Thermal neutrons, 174
Thermoelectricity, 95
Thermonuclear reaction, 207
Thomson, George P., 102
Thomson, Joseph J., 37, 54, 57, 112, 138
Thorium, 110
Thorium-*231,* 137
Thorium-*232,* 126-128, 150, 205
half-life of, 137; series of, 130-131.
Thorium-*234,* 126, 128
Thorium series, 130-131
Threshold value, 55
Tin, 172, 187
Townes, Charles H., 98-99
Tracers, isotopic, 165
Transistors, 95
Transition elements, 83ff.
Transuranium elements, 178
Tritium, 169
Triton, 169

Ultraviolet radiation, 33-34, 209
Uncertainty, principle of, 107, 243-244
Universe, 231-233
Uranium, 109-110
earth's age and, 150-151; energy produced by, 189-190; fission of, 192-194; neutron bombardment of, 176.
Uranium-*233,* 204-205
Uranium-*235,* 131-132
fission of, 197-202; half-life of, 136-137; series of, 131-132.
Uranium-*238,* 150, 197-198
breakdown of, 126, 183-184, 189-194; fission half-life of, 194; half-life of, 135-136;

packing fraction of, 187; series of, 128-129.
Uranium hexafluoride, 198
Uranium series, 128-129
Urbain, Georges, 65
Urey, Harold C., 144

Vacuum, 29-31
Vacuum tube, 43
Van de Graaf, Robert J., 158
Veksler, Vladimir I., 160
Virtual particles, 244
Volta, Alessandro, 53
Voltage multiplier, 158

Walton, Ernest T. S., 157
Water, 11
molecular weight of, 19
Watson-Watt, Robert A., 51-52
Wave mechanics, 105
Weak interactions, 248
Wien, Wilhelm, 112
Wigner, Eugene P., 196, 255
Wilson, Charles T. R., 116, 217
Wu, Chien-Shiung, 255

Xi particles, 252-253
X ray(s), 61ff.
burns from, 209; characteristic, 62-65; discovery of, 34-36; muons and, 246; production of, 61-62; radioactivity and, 108-111; tracks of, 117; wavelength of, 63.
X-ray tube, 62

Yang, Chen Ning, 254
Yukawa, Hideki, 243-247

Zeno, 3
Zinc, 188
Zinc sulfide, 153
Zworykin, Vladimir K., 50

Other MENTOR and SIGNET Science Titles You'll Want to Read